Thanks to my patient wife,
who makes my life so enjoyable

I am thankful for—

My son Levi, with his encouragement to use sound, gracious words

My dad, who, although believing Earth is young, gave a lot of his time to provide invaluable critique and input

Dr. James Seaborn, Professor of Physics, University of Richmond, for his insights and kind suggestions

Dr. Sean Agnew, Professor of Materials Science and Engineering, University of Virginia, for his helpful and meticulous review

Dr. Mary Schweitzer, Professor of Biological Sciences, North Carolina State University, paleontologist and believer, for her encouraging words and setting an example of trying to disprove what we suspect to be true

1

Science and the Bible

During a dark time of scientific repression by Christian orthodoxy[1], Galileo Galilei bravely published his book in 1632 explaining how Earth orbits around the sun. But the ruling church of his time believed it was an absolute fact of scripture that the sun rises and goes down— Didn't the Lord say the sun rises and goes down? Didn't David say in Psalms that God laid the foundations of the earth so it should not be moved forever? Doesn't Earth have four corners?

Galileo threatened their long-held beliefs on something they could not *possibly* be wrong about. They enforced their dogma by condemning Galileo of heliocentric heresy, declaring his book was "false and contrary to the Holy and Divine Scriptures."[2] (Psalms 50:1, 104:5; Malachi 1:11)

Galileo didn't have access to the knowledge we now have to consider if Earth's four corners and four winds refer to Earth's magnetosphere and four-part relationship with the solar wind magnetopause, magnetic reconnection, and the north and south poles.[3,4,5]
(Isaiah 11:12; Revelation 7:1, 20:8)

Under the threat of torture and death, Galileo recanted his findings to the ruling organization of his time and place, so they relented and just sentenced Galileo to a lifetime of house arrest.[6]

If you could see the earth illuminated
when you were in a place as dark as night,
it would look to you more splendid than the moon
- Galileo Galilei

Just because opinions are firmly enforced for long times by powerful people of influence doesn't make them true.

Some Christians agree that recent scientific findings, such as neutral mutation[7] and radioactive testings,[8] indicate that modern humans have been around for over one-hundred thousand years.[9] But some Christians argue it is an absolute fact of scripture that Earth was made in six solar days about six thousand years ago.[10,11,12]

Is the strong recent push for six solar days of creation to counteract the morally void theories that evolutionists teach in public schools, as if they were facts? Of how over billions of years dinosaurs evolved from muck, primeval species mutated into new and improved species, and of how, after many new bursts of life, a chimpanzee-type species eventually mutated via natural selection and small random/deterministic drifts of genetic changes into modern humans about 200 thousand years ago?[13,14]

These theories threatened my core beliefs, because my belief in six solar days of creation was connected to the Bible's validity, and to Jesus. I was ensnared with discounting sound science, because I thought it proved evolutionists were right and the Bible wrong. I didn't consider if it was my *understanding* of Genesis that was in error, and that evolutionists were also mistaken about *their* conclusions from scientific evidence.

Evolution has no purpose – Yuval Harari

Could it be that evolutionary theory *and* young earth creationists are *both* wrong? If God authored the DNA code for these new bursts of life during the last 4 billion years or so, then wouldn't this explain why paleontologists have not found sound evidence for the billions of missing links and intermediary gaps that should exist if these creatures had indeed mutated? Are these links "missing" or simply "nonexistent"?[17]

What if the first five chapters of Genesis are literally written verbatim in the same chronological order of time as the events occurred? What if Genesis 1 tells of modern humans being created during the sixth time period, and Genesis 2 explains how Adam was made directly by God as a son of God, about seven thousand years ago? If God did make Adam *after* the humans of Genesis 1, what are the theological implications?

Sound science wasn't a bane of Christianity over a hundred years ago. Sound scientific findings should *strengthen* a believer's faith. Respected Christian leaders such as Cyrus Scofield (*Scofield Reference Bible*), Clarence Larkin (*Dispensational Truth*), George Pember (*Earth's Earliest Ages*), Simon Episcopius and Thomas Chalmers all thought that there may have been an Era older than ours, and that Earth might be much older than six thousand years. Simon Episcopius was preaching of an Era before ours nearly two hundred years before the first scientific discovery of prior Eras by Georges Culvier, the founding father of paleontology. Culvier was hotly disputing evolution some twenty years before Darwin even wrote his book.

I do not think it is necessary to believe that the same God who has given us our senses, reason, and intelligence wished us to abandon their use, giving us by some other means the information that we could gain through them
- Galileo Galilei

Would the authors of Genesis be incredulous that many people now think Adam and Eve were created on the sixth day, since it was so plainly written that they were made afterwards? They might be irate with Stephen Langton for twisting their words with his chapter and verse placement, pointing out that, since the rest of Genesis was written in the same order as the events occurred, then how could anyone possibly assume the making of Adam and Eve was elaborating upon the sixth day?

Did proving Earth rotates around the Sun disprove the Bible? At some point, religious folk of the past had to stop taking comfort in their complacent religious traditions. Wisdom is realizing we are ignorant. Worrying about how truth may negatively impact our relationships with loved ones is not trusting in God. Truth gains honor and value over time. (Proverbs 4:8, 15:33, 30:2; Zephaniah 1:6)

God gave us a simple and accurate depiction of how He created the Universe and Earth to an ancient people who were busy obtaining food and water and fighting and loving, much as we do now. He understood our perspective and just told us what we need to know. Genesis wasn't written as a textbook pre-rebuttal to evolution. The Bible was written for the far more important reason of leading human souls to Christ.

Genesis gives a broad overview of our planet's origins to a people who knew nothing of any Triassic, Jurassic and Cretaceous periods. They watched the sun rise and set. They saw Earth as a circle all around them on the horizon when they were high on a mountain top, or far out at sea. Just climbing a mountain or sailing into open sea was difficult then.

Scripture is a book about going to Heaven.
It's not a book about how the heavens go
- Galileo Galilei

When a sound scientific finding conflicts with how we interpret a passage, God wants us to re-evaluate our views, not cling to long-held ideas of limited human wisdom that aren't supported by scripture, or simply ignore what is plainly stated.

It *is* powerful evidence that the Genesis 1 sequence, written long before sound scientific findings, is in perfect scientific sequence, and that the Big Bang theory at long last essentially proves, or at least supports, that God created the Universe and is stretching it out. It's as if Science reversed engineered Genesis, and now claims it as their own.

God obviously wants us humans to pursue science. He purposely gave us the abilities and desires to uncover His tantalizing scientific mysteries, such as perching Earth in a special place in the Milky Way so that we have a fantastic view of the Universe. He made the sun 400 times larger and 400 times further away from the moon to make possible the unique scientific discoveries that arise from our near-perfect solar eclipses, such as in 1919 when Einstein's theory of general relativity was confirmed.

Earth is in a perfect galaxy to sustain life, perfectly positioned safely away from the cosmic explosions near the massive black hole at the Milky Way's center, yet close enough to gain *some* heavy elements via supernovas, and yet safely out of reach of the hazardous Sagittarius and Perseus spiral arms. (Isaiah 48:13; Job 38:4; Psalms 102:25, 104:5-6)

He stretches out the north upon empty space
and hangs Earth on nothing!
- Job 26:7

The Spherical Earth and Eternal Universe

Galileo understood that the Bible, although true, was written to the perspective of those living long ago and is subject to the bias of its translators and expounders. When our spirit detects our imprinting might be biased and in error, our soul can react in fear by quickly and intensely justifying our errant views. Those who opposed Galileo could not reconcile astrophysics with the Bible. They may have feared risking an empty void of unbelief if they *did* try to understand it.

Some ancient astronomers did think Earth rotates around the sun, but they also observed the predictable rotation of the stars across the night sky and concluded that our Milky Way Galaxy encompasses the entire Universe and had existed forever. Most scientists agreed with this eternal universe assessment. But Christians insisted the Universe had a beginning, and it began when God created it.

The Eternal Universe Theory was widely accepted by the scientific community until it was shattered in 1924 with Edwin Hubble's discovery that some stars are rapidly retreating galaxies! It is now widely accepted among scientists that the entire Universe sprang into existence nearly 14 billion years ago, and is still rapidly expanding.

Humanity has inherently assumed the Universe revolves around us, and there are no other creatures of much significance. But the evidence indicates that we are just a tiny sliver of what God has done in past eons. The Bible's account of modern humanity from Adam to Jesus is a miniscule part of what God has done in the past, and of what God will do in the future. God offers us hints, not details, of the distant past and future. The Bible is about God dying in human form to save human souls and spirits, which wasn't done for angels and the other creatures in the heavenly realm. Perhaps God places so much value on us because we are made in His image. (John 3:16; Romans 8:17; 1 Peter 1:12)

Such is the grasping tendency of the human heart,
that it must have something to lay hold of and which,
if wrested away without the substitution of another something in its place,
would leave a void and a vacancy as painful to the mind,
as hunger is to the natural system
– Thomas Chalmers

Evolution

If God didn't determine our genetic processes, then what force did?

Neo-Darwinistic and Modern Synthesis evolution portray an unintelligent-yet-astonishing deterministic god-like creative force to account for our incredible complexity of life. But, where are the compelling explanations for those evolutionary "bursts of life gaps", the mysteriously missing billions of intermediate complex life form links that should exist to support these alleged mystical bursts of mass mutations?

Darwin said, "If it could be demonstrated that any complex organ existed, which could not possibly have been formed by numerous, successive, slight modifications, my theory would absolutely break down."[16] Some modifications do exist,[17] but shouldn't the evidence for the billions of intermediary links be obviously compelling and prolific? Some intermediaries are offered,[18] but believing it all randomly drifted via small genetic changes into a vast spectrum of living things requires a bit more than being firmly planted in the soil of scientific self-knowledge. Scientists of *www.thethirdwayofevolution.com* think Neo-Darwinists have elevated Natural Selection into a unique creative force to solve all difficult evolutionary problems without a real empirical basis, and alternatives to traditional evolutionary theories should be considered. Science evolves.

Some professionals state on *www.dissentfromdarwin.org*, "We are skeptical of claims for the ability of random mutation and natural selection to account for the complexity of life. Careful examination of the evidence for Darwinian theory should be considered." They hold doctorates from Oxford, Cambridge, Harvard, Dartmouth, Rutgers, University of Chicago, Stanford, and the University of California at Berkeley. Many are professors or researchers at major universities and research institutions such as Cambridge, Princeton, MIT, UCLA, University of Pennsylvania, University of Georgia, Tulane, and in Moscow, Japan and Israel.

Every tiny detail of any living creature is revealed to be an incredible work of art under an electron microscope. Fascinating details emerge about their behavior, symbiotic relationships in how they relate to and help other creatures and plants, their chemical reactions, DNA, unique chromosome strands, molecular engines, and trillions of cells working together in incomprehensible unity.[19]

Consider the modern medicines that have been bio-emulated from plants, mold, leeches and tree bark, and the thousands of medicines yet to be discovered. Are they miraculous freaks of nature? Consider a teaspoon of sea water has more living organisms than the entire planet's human population, and a teaspoon of fertile soil has a billion inhabitants. Consider the billions of neurons forming a brain in the head and tentacles of an octopus that enable it to change color and sneak out at night to snack on crabs in other aquariums, or the 100k megapixel skin of a cuttlefish with a vastly higher resolution than an 8k TV.[19]

Many biologists think that 99 percent of our planet's total species lived during the last five Eras and are now extinct.[20] Have all of these billions of animal species, insects, microbes, and plants *really* been simultaneously, and very prolifically, mutating into other species during the last 4 billion years?[21] If so, where is the conclusive evidence?

Six billion nucleotide bases make up human chromosomes and are in *each* of our ten trillion or more cells. Billions of chemical reactions occur each millisecond in every living creature, billions of proteins message, chemically switch off and on, give electrical charges, and it must all be in near-perfect working order for each creature to stay alive. The tiniest creature has millions of components working together in perfect unison, *and* it can eat and reproduce.[19] Did this *really* all happen by very determined and magical coincidence? Imagine natural forces evolving sand into silicone and then into a smart phone chip with only a million circuits or so, along with the towers and network it needs to function. But tiny organisms are far more complex than smart phones, and require many other life forms to function and exist. All living things depend on the lives of others. DNA code is vastly more advanced and intricate than a simple smart phone that is unable to even reproduce itself.

Ask the animals and they will teach you
The birds will tell you
Speak to the earth and it will instruct you
Yes, those creatures of the sea
Who does not know the hand of the Lord made them?
The souls of all living things are in His hand
and the breath of all humanity
Job 12:7-10

The DNA Code

What if the DNA of Earth's living creatures all share common evidence of God's authorship? Was it evolution or God who authored DNA code? Dr. Geoffrey Simmons, in his book, *What Darwin didn't know,* explains from a fascinating medical viewpoint the improbability of DNA and chromosomes evolving to create the highly intricate genetic code, chemicals, biological systems and emotions of a human—

For example, if the alleged 6-million-year-old common ancestor of chimpanzees and humans had a brain size of 30 billion neurons, then modern humans would need to mutate 60 billion more neurons in 6 million years to achieve our current 90 billion neurons! This averages about 100,000 neurons per generation, but evolutionary theory claims it occurs in evolutionary bursts, so some lucky being would need to mutate 1-2 million more neurons every 1,000 generations.[22]

Simmons points out that our vast genetic variances and biochemical activities formidably challenge deterministic genetic drift, requiring an astounding 133 or so predetermined genomic mutation improvements per generation to evolve from a chimpanzee-like creature into the 6-billion-bit human. We have a better chance of winning the lottery every week, because it is very rare to evolve even a small insignificant spontaneous evolutionary change in 100 years, and requires DNA lottery-winning parents to have offspring with the same "defect", and have that defect be an extremely rare spontaneous improvement, and not the usual detriment, which also continues to be passed on to their descendants, so that in a mere 4.7 billion years this magically accounts for the vast chemical and molecular complexities of DNA, smell, taste, sight, nerve, muscle and organ functions for our vast millions of living creatures, and also somehow accounts for emotions, humor and art.

The consciousness of man, the mind, is something not to be reduced to brain mechanisms - Wilder Penfield, father of modern neurosurgery

Bacteria and insects *can* adapt by altering their DNA with random genetic mutation, and natural selection enables the survival of their fittest variations. Bacteria can even transfer genes and swap DNA with other bacteria, but, were they designed to do that?[22,23] Where is the strong evidence for billions of mammals adapting by wide-spread beneficial mutations? Does it take faith to conclude all living organisms evolved?

We logically assume there is intelligence behind the design of any computer or machine, so is it not extremely speculative to argue an unintelligent force created these vastly more complicated creatures?

Robert Grass, Ph.D. of ETH Zurich, said, "A little after the discovery of the double helix architecture of DNA, people figured out that the coding language of nature is very similar to the binary language we use in computers. On a hard drive, we use 0s and 1s to represent data, and in DNA, we have four nucleotides, A, C, T and G."[24]

DNA is massively superior in storing information than our current computer technology. The total of all the world's information, about 2 zettabytes, could not only fit on a four-gram DNA hard drive the size of a teaspoon, but would also last for thousands of years![25]

DNA is like a computer program, but far, far more advanced than any software we've ever created - Bill Gates

Teams of engineers, corporations and governments have made great efforts in advancing our computer technology, and some are now developing DNA-based technology for storage.[25] Did this vastly more advanced DNA technology *really* just *happen* to come into existence?

The DNA code of Earth's living creatures could only have been authored by the same unimaginably intelligent Creator. Our planet's ecosystem is an absolutely brilliant masterpiece.

After reviewing and studying the neutral mutations of millions of planet-wide DNA tests from 100,000 different species, two scientists found that 90 percent of our current animal species, including humans, all emerged nearly simultaneously between 100,000 to 200,000 years ago. They also found that our current animal species have very clear genetic separation from each other.[7] This scientific evidence indicates that our current species are *not* genetically drifting. This also ironically confirms the original findings of Georges Cuvier, the renowned founding father of paleontology, who was convinced our current species are not evolving.

Scientists have no proof that life was not the result of an act of creation,
but they are driven by the nature of their profession
to seek explanations for the origin of life
that lie within the boundaries of natural law
– Robert Jastrow, of NASA

Probability

Does 4.7 billion years allow enough time for Earth's billions of creatures to have evolved via genetic drifting? Evolutionists need trillions of years and unproven multi-verse universes to make it more probable. Should adding more zeroes in the years ago allow evolutionary processes to take the credit for our existence?

Some biblical prophecies of Jesus were accredited by the American Scientific Affiliation— the odds of Jesus fulfilling eight prophecies has the same odds as covering the state of Texas two feet thick in silver dollars, marking one of them, and then a blindfolded person randomly picks that marked silver dollar. The number of silver dollars needed to cover Texas two feet thick is 10^{17} (10 with 17 zeros after it).[26]

The odds of Jesus fulfilling 48 known prophecies is 10^{157}. Instead of Texas covered in silver dollars, imagine randomly selecting one single marked electron from the entire Universe. But, since the observable Universe only has about 10^{80} electrons, we would need more universes to pick our marked electron from (electrons are so small it would take billions of years to count just one cubic inch of them). To match the odds of Jesus fulfilling 456 known prophecies, we would need even more universes to pick this single marked electron from. In other words, if the Gospel accounts are true, and the Old Testament prophesies predicted a Messiah, then Jesus is very likely the Son of God.

Major movies employ intelligent people to craft believable stories, but they still leave plot holes in their fabricated storylines. In comparison, the Bible's vastly greater storyline aligns nearly perfectly, yet was written over thousands of years by many people who never knew each other.

After investigating some Bible topics I thought had inconsistencies, I found truth and harmony, so I believe in the infallibility of the Bible as it was originally written from my own personal studies and experiences. Josh McDowell set out to disprove the Bible, but discovered how credible and historically accurate it is, and became a believer. His book *Evidence that Demands a Verdict* details his findings.

It has been said that there is nothing more uncommon
than common sense
– Thomas Chalmers, Church of Scotland, 1836

Is the true meaning of Genesis misunderstood? Does Genesis 1 explain that the humans created during the sixth time period were living long before Adam was made in Genesis 2? If so, why did God do that?

Has God created six Eras of living creatures and dinosaurs, changing each of the first five Eras with transitional extinction events so that newly created living creatures can flourish?[27]

God made our current Era with a mass extinction event planned for its end, and He will then make a new and improved Era to take its place. If it is possible that our current Era is God's sixth, then would it not be presumptuous to argue that God's other Eras never existed?

In questions of science, the authority of a thousand is not worth the humble reasoning of a single individual - Galileo Galilei

I've learned not to throw breadcrumbs overhead to flying seagulls. It is much better to toss them out in front so the crumbs land in the sand. Have you ever noticed how, if there are just a few seagulls, they are very cautious and keep their distance, but if there are many seagulls, they all become bolder and more self-confident, and edge closer and closer? I've noticed people behaving this way with wild buffalo in Yellowstone. If there are just a few people then they keep their distance, but if many people are crowding around to view the buffalo, then some boldly edge closer and closer to the buffalo. The threat to the seagulls, or to the threat to the humans, doesn't change, just their perception of the threat. There is a false feeling of safety in the collective illusion of groupthink. The same is true within the cultures of religion and evolution.[28] It is simply more comfortable to safely view things from the same perspective as those around us, and try not to think about it all that much.

It helped me to understand two very different perspectives by reading two inspiring spirit-filled books by Christian author Bob Goff, *Love Does: Discover a Secretly Incredible Life in an Ordinary World*, and *Everybody, Always: Becoming Love in a World Full of Setbacks and Difficult People*. Then, from a starkly different perspective, reading renowned atheist and evolutionist Richard Dawkins' books, *The God Delusion* and *The Greatest Show on Earth*. Richard Dawkins was raised in a Christian home.

If sound science appears to contradict the Bible,
we may be sure that it is our interpretation of the Bible that is at fault
Christian Observer (1832)

2

Genesis Inconsistencies

When searching for the truth, it is essential for an investigator to avoid preconceived bias, foregone conclusions, and tunnel vision on any one facet, and to keep digging until the truth is uncovered. Every door needs to be opened to see what evidence lies beyond. Specialists in astrophysics, archaeology, anthropology, geology, theology, zoology, paleontology, medicine, history and linguistics can provide invaluable knowledge and insights, but tunnel vision can arise from a narrow focus on a single specialty. Also, some inherent indoctrinations are encouraged in each specialty that tends to filter the illuminating streams of thought that enlighten the broad insights of minds open to many possibilities.

I am well aware that because I did not study the ancients, some foolish men will accuse me of being uneducated. They will say that because I did not learn from their books I am unqualified to express an opinion. But I would reply that my conclusions are drawn from firsthand experience, unlike the scholars who only believe what they read in books written by others
– Leonardo Da Vinci

Some Christians believe human origin began about 6 thousand years ago, but some have discounted Genesis, believing it must be poetry or abstract,[29,30] since scientific and archaeological findings clearly show human remains and activity date back to about 150 thousand years ago.[7,8] Atheists and Humanists argue that this 6-thousand-year creation view is solid evidence that the Bible is false.[1,28.31,32]

Albert Einstein said in his autobiography that his deep religiousness– "reached an abrupt end at the age of twelve. Through the reading of popular scientific books, I soon reached the conviction that much in the stories of the Bible could not be true… it was a crushing impression.

Mistrust of every kind of authority grew out of this experience, a skeptical attitude toward the convictions that were alive in any specific social environment." If Einstein had realized Genesis supports widely accepted mainstream conclusions arising from astronomical evidence, and recent neutral mutation DNA and radioactive testings, would he have considered his discarding of the *Bible* as his greatest blunder?[7,9,46,77]

Understanding the first five chapters of Genesis as written verbatim in the same order as the events occurred gives a fascinating glimpse into the past, and also confirms paleontological and anthropological findings of the six major Eras. For example, God made Adam to cultivate and be a farmer, so is it just a coincidence that his arrival intervenes with the advent of modern agricultural canals and crop rotation, the abrupt appearance of cows, goats, horses and sheep, and Earth's very first modern cities? It was timely, since those increasingly dense populations had dire needs for them.

For thousands of years some religious leaders believed it was an absolute fact of scripture that the sun moved around the earth, since the Lord clearly says that the sun rises and goes down. (Malachi 1:11) Of course, we now know that they were wrong and Galileo right, but a spherical earth rotating around the Sun hasn't altered the validity of the Gospel. Arguing that the sun rotates around Earth would now be regarded as ignorant.[34]

Genesis was written so that people who knew little of science could understand it, but this doesn't mean we should be ignorantly critical of God's scientific processes! Understanding sound scientific discoveries and how they impact scripture increases one's faith in God. Truth leads us to God. What does ignorance lead to? Learning more of astrophysics made me realize how our incredible God uses supernovas. What I will be proposing is not a threat to the Gospel of Jesus. It is not the Bible that is in question, but our indoctrinations that can warp our interpretation of it.

The main question is whether people know
the truth about themselves
- Yuval Harari

Twelve Genesis inconsistencies arise from traditional theological views of Genesis. Atheists and Humanists argue that the first two chapters of Genesis are inconsistent and therefore untrue.[31]

1. Did God create the heavens and the earth, *then* make light, *then* make day and night, and *then* make the sun, moon and stars?

2. Why does Genesis 1 state that plants were created first, animals second, and humans third; but in Genesis 2 Adam was made first, then trees are planted, then farm animals made, and then Eve?

3. On the sixth day, were Adam and plants and animals made, animals named, no comparable helper found for Adam, Eve made, and *then* God blessed them and told them to multiply, *all* in less than 12 hours?

4. Did God *really* make Adam with an emotional need to love a woman and be able to reproduce, but then bring farm animals before Adam to see if they could possibly work out as a counterpart helper for Adam?

5. Why did God stop creating after Genesis 1:1 and just "let" things become something, until He resumed creating creatures on the fifth day?

6. Why were humans on the sixth day "created," but Adam and Eve "formed" in Genesis 2?

7. After murdering his only brother, who did Cain immediately worry would kill him, if Adam and Eve were the only other living humans?

8. How could Cain become a fugitive if it was just Adam and Eve who existed at the time?

9. Who did Cain build a city for?

10. Who did Cain and Seth marry? If it was incest, is that consistent with God's moral laws? What made it wrong for Cain to murder Able?

11. Why does Genesis 5 state this book is about Adam's genealogy? Who else could it be about if Adam was the only human made by God?

12. Who are the men in Genesis 4:26 that began to call on the name of the Lord? Adam, Cain, Lamech, Seth and Enos already knew the Lord.

If the Holy Spirit is nudging you towards this subject, or if you sense that evolution requires a bit more faith than you have for it, then these following chapters are for you. The truth *is* important, isn't it? But truth cannot be found in bending the Bible to match one's understanding of science, or in bending science to match one's understanding of the Bible.

The Genesis Synopsis

Viewing Genesis from the perspective of those living long ago in Mesopotamia, and understanding it was written in the same order as the events occurred, resolves these twelve inconsistencies and also confirms sound scientific findings—

God first created the Universe and planet Earth and then made six major Eras of amazing living creatures on Earth. God instantaneously created some new bursts of life for each Era, and may have genetically modified some existing creatures into new intermediaries.

The first five Eras ended with transitional mass extinction events, to change each old Era into a foundation for the next new Era of God's newly made creatures to flourish in. These first five major Eras are composed in the fossilized layers of Earth, and are the visible folded garments of God. Ironically, the areas where the garden of Eden is believed to have been rests on top of sedimentary rock composed of the Tertiary, Cretaceous, Jurassic, Triassic and Paleozoic layers, and all the way down to the Precambrian basement.[35]

Earth was left as a stormy ruin at the end of the fifth Mesozoic Era. The making of our sixth Era begins in Genesis 1:2 and gives an accurate, but very broad, overview of God's sequence of events— God initializing life first in the oceans, the womb of the earth, to make a new atmosphere and thin the dense planetary cloud cover enough so some sunlight shines through, lets the oceans recede back to expose land, makes plants to modify our atmosphere even more and clear the cloud cover so the sun, moon and the stars become visible again from Earth's surface, then creates sea creatures and birds, makes more mammals and reptiles, and then finally creates humans.

God blessed these modern humans created during the sixth time event, perhaps about 150 thousand years ago, telling them to fill and subdue Earth. They diligently populate and spread throughout the planet, causing the extinctions of megafaunal animal species. These intelligent humans may have had no knowledge of good and evil, and may have been long-lived and more intelligent, innocent and socially adept than dolphins, whom God had created, gave souls or life to, and blessed on the fifth day.[33]

About ten thousand years ago, as the melting glaciers of the Ice Age retreated, the climate stabilized and land began to dry. The increasing human population was retracted into small areas near rivers.[37]

These modern Genesis 1 humans were having a difficult time managing their dense populations. They were proficient at reproducing and spreading all over Earth to find tree and plant seeds to eat, and killing the giant megafaunal, and but were not so good at building sophisticated canals and coordinating large cities. There were no modern farm animals, no methodical and sorely needed controlled canal systems for irrigation, no crop rotation methods, and no coordinated government to support the growing populations, such as in Uruk and Ur.[38,39,40,41,42,43]

God, of course, knew of these dire human needs. So, about 144 thousand years *after* creating the first humans on the sixth day in Genesis 1, God notes help is needed to facilitate their changing environment, and directly makes a new modified human, Adam, a son of God, with the mental ingenuity to assist these Genesis 1 humans. God then plants a garden in Eden with modified agricultural plants and trees, and places Adam there to care for them. God makes new agricultural animals to help Adam, and then makes Eve, since none of the existing women are found to be an adequate counterpart for this new modified human male.

God tells Adam he can eat fruit from the Tree of Life, but not the Tree of Knowledge of good and evil, or else he will die. Adam disobeys God, and so Satan, also a son of God, gains Earth's power of death and the key of Hades. (Job 1:6-12, 2:1; Hebrews 2:14; I Corinthians 15:55-56; Ephesians 4:7-10; Revelation 1:18) (Detailed in the *God's Master Plan* chapters "The Conflict in the Heavens," "The Book of Life" and "The Spirit and Super Body")

God then tells Adam and Eve they will reproduce, knowing Jesus, a son of God, will descend from Adam and defeat Satan, so the now errant humans of Earth, and those in the heavens, can be reconciled to God. The Bible tells of how Adam's descendants led to Jesus, and how God helped humanity by introducing His superior moral codes, some of whom fell into utterly barbaric practices. These moral codes form the bedrock of, and still greatly influence, our modern civilizations, from the Sumerians and forward. The Bible explains how to keep our heart and care for our spirit, and how to obtain everlasting life. (Colossians 1:15-20; Revelation 12:12)

How foolish can you be?
He is the Potter and certainly greater than you, the clay!
Should the created thing say of the one who made it, "He didn't make me"?
Does a jar ever say, "The potter who made me is stupid"?
- Isaiah 29:16 NLT

Beliefs vs. Views

Belief is a mental acceptance that something is true and exists. Belief is placing trust and faith in something, or someone. It often begins when we are young. A belief is a deeply felt conviction within our spirit that becomes deeply rooted in what we think to really be true. It alters our lives and influences the decisions we make, such as believing in Jesus and that He is the Son of God who died for us. And so we become a believer.

Views are opinions, a judgment call on something we think *might* be true, but we are not quite sure about, something we aren't quite willing to wholly accept as a truth or belief. A view is a logical perspective, a decision that is held with varying degrees of confidence but has not been substantiated with positive knowledge or proof. It could be going along with a prevailing view that many other people have but that we haven't ever looked into for ourselves, such as considering ourselves a Christian but never actually believing in Jesus. A view is a conclusion on something that may have been thought about or even researched, but it is not quite important enough to accept as one of our firmly held beliefs.

The danger in making something a Belief that should just be a View is that we tend to attach this belief to other firmly held beliefs. For example, if a believer in Jesus also believes deep within their heart that Earth is certainly no older than six thousand years, then being confronted with new information that casts doubt on their young earth belief can also shake their foundational beliefs of the Bible's validity, make them defensive, and even emotional, since this casts doubt on Jesus being God's Son. It's fear.

The danger of the snare in assuming that a particular view of creation should be a doctrine that is synonymous with the gospel of Jesus is that one's vital faith and relationship with God is conditionally connected to something of dispute and uncertainty.[9,10,11,44] One's faith is then vulnerable, and one's conscience can weaken. (1 Corinthians 8:9)

These Bible-based <u>views</u> of Genesis are presented with the intention of strengthening our faith in God, to demonstrate that God's Word truly is amazing, and to also encourage those who have been ensnared by unintentionally allowing a View to prematurely develop into a Belief.

Knowledge puffs up, but love edifies. (1 Corinthians 8:1)

Facts are Stubborn Things

The Bible was written to the perspective of ancient humans living thousands of years ago so that they could understand what God said. The Lord said the sun rises and goes down, and we observe that it still does. (Malachi 1:11)

But now we see satellite images of Earth, and learn of fossils, DNA, atoms and supernovas. We know more than ancient civilizations, but we tend to assume that they shared this same knowledge and understanding we now have, or we judge God and think that He should have explained His scientific principles to them, and to us, in the Bible. Although the Bible is not a textbook on anthropology, astrophysics or geology, it does offer brief explanations of science. (Job 26:7-8; Jeremiah 33:22; 1 Corinthians 15:41)

Humanity has historically attributed the scientific mysteries of our world to God. But as scientists uncover how God's precise physical formulas and genetic codes work, these discoveries are often presented as evidence that there is no God, since humanity (sort of) figured out *how* it was done. But finding out *how* God did it doesn't eliminate God! (Proverbs 30:18,19)

God gave humans the desire and ability to uncover His scientific mysteries and past Eras, and these discoveries shouldn't be disregarded by believers. We need to evaluate how sound scientific findings impact our views of certain passages, even if it conflicts with our current assessment of what we traditionally think they may mean. (Acts 17:11)

The broader question is *why* God did it. When I've had concerns about the Bible's validity, and after closely researching these inconsistencies, I have found meaningful understanding which has convinced me even further that the Bible really is true and accurate. The Genesis account began to clarify after I finally admitted my past understanding was clouded, even after inventing additional things to try and make it all work. After investigating and examining the evidence and clues, I discovered Genesis validates sound scientific findings, freeing my mind to reflect on *why*.

With my Young Earth upbringing, I expected to quickly and easily disprove this thought that humans of the sixth day in Genesis 1 were created well before Adam was made in Genesis 2. But after digging deeper, I discovered more evidence matching what I at first only suspected might be true, however strange and bizarre it seemed to me at the time.

If one had to guess what other planets in our solar system are like, without knowing anything about them, who could correctly guess their greatly varied sizes, shapes and characteristics? Truth is often stranger than fiction, and much more interesting. (1 Corinthians 15:41)

This view of humans living before Adam adds depth to the gospel of Jesus being the son of God and dying for our sins by highlighting *why* Jesus had to descend from Adam, and it reveals more of God's mercy and compassion for humanity. God is a caring God of justice and order, but most importantly, God is love. It reveals more of how complex God is, how little we know of our own Era, gives glimpses into God's prior Eras and the current conflict in Heaven, the Book of Life, human souls, spirits and superbodies, and humanities future. It shows our relationship with the Lord is vastly more important than our relationship to Adam.

Believers who are scientists seek to discover *how* God did it, and tend to bend scripture to match their current understanding of science. Theologians seek to understand *why* God did it and tend to bend science to match their current understanding of scripture. This investigation seeks to uncover the truth, and with all diligence to not bend either science or scripture, nor misrepresent *any* person or group. My sincere apology is offered in advance if I have inadvertently done this!

We learn by trusting in God and sincerely desiring to seek the truth, not by becoming mired within our own opinions. Some Christians cling so tightly to their cherished ideologies that they never consider how their biased beliefs might be turning others away from Christ. Sound science reveals God's truths and turns humans towards Christ.

Consider the old earth views of Cyrus Scofield (*Scofield Reference Bible*), Clarence Larkin (*Dispensational Truth*), George Pember (*Earth's Earliest Ages*), Simon Episcopius, Thomas Chalmers and Georges Culvier (the founding father of paleontology). Simon Episcopius was preaching of an Era before ours nearly two hundred years before Georges Culvier's scientific discovery of prior Eras. Sound scientific findings should *strengthen* a believer's faith.

Facts are stubborn things;
and whatever may be our wishes, our inclinations,
or the dictates of our passion, they cannot alter
the state of facts and evidence
– John Adams

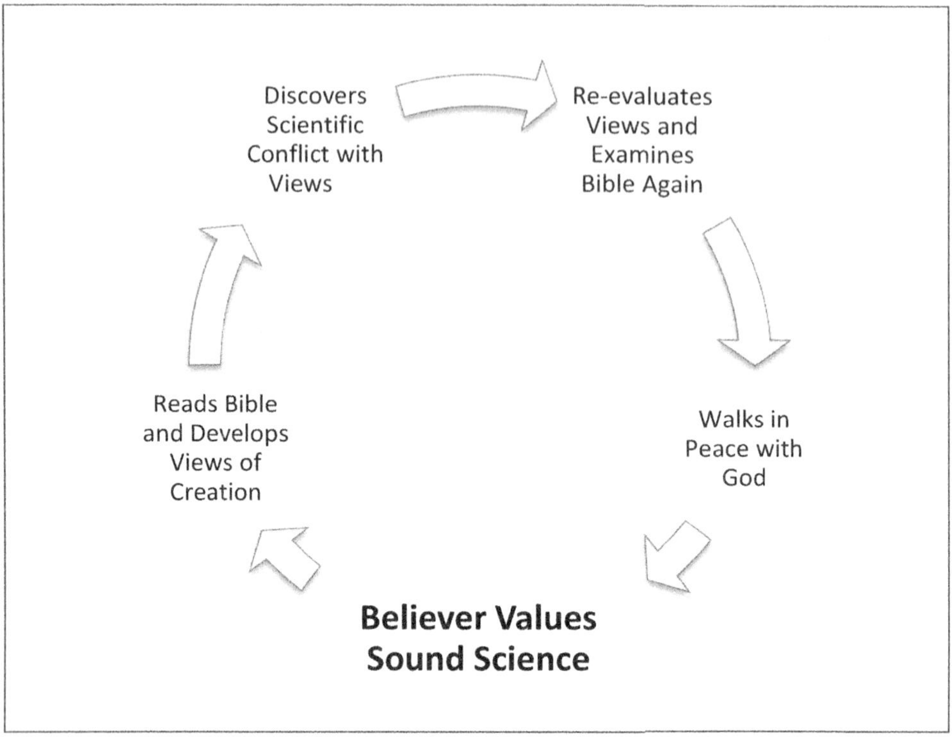

Test everything that is said. Hold on to what is good - 1 Thessalonians 5:19 NLT

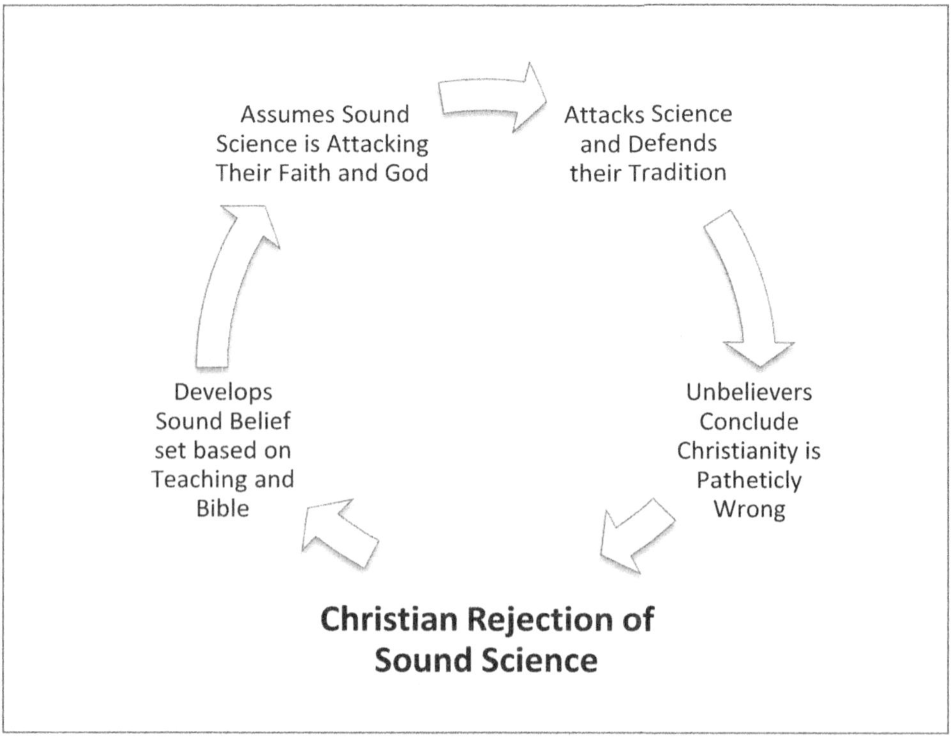

3

Genesis 1

God's Garments

God's divine nature and eternal power is clearly visible here on Earth and in our skies. (Psalms 19:1-3) Our planet's cloak of creation is God's icon. (Romans 1:20) God's iconic craftmanship from past creations is revealed in the fossilized layers of Earth, which are His folded garments—

"Of old You laid the foundation of the earth, and the heavens are the work of Your hands. They will perish, but You will endure; yes, they will all grow old like a garment; like a cloak You will change them, and they will be changed." - Psalms 102:25-26

"You, Lord, in the beginning laid the foundation of the earth, and the heavens are the work of Your hands. They will perish, but You remain; and they will all grow old like a garment; Like a cloak You will fold them up, and they will be changed. But You are the same, and Your years will not fail." - Hebrews 1:10-12

"Lift up your eyes to the heavens, and look on the earth beneath. For the heavens will vanish away like smoke, the earth will grow old like a garment, and those who dwell in it will die in like manner; but My salvation will be forever, and My righteousness will not be abolished."
- Isaiah 51:6

"Have you commanded the morning since your days began, and caused the dawn to know its place, that it might take hold of the ends of the earth, and the wicked be shaken out of it? It takes on form like clay under a seal and stands out like a garment. From the wicked their light is withheld, and the upraised arm is broken." - Job 38:12-15

"They" and "them" and "all" are plural— "they will all grow old like a garment; Like a cloak You will fold them up, and they will be changed." God's wisdom is often multifaceted, and this analogy of God having multiple cloaks of creation has two aspects—

The first aspect compares our current Era's life forms to the cloak of creation that God is now wearing. God will change His cloak of creation at the end of our Era with a mass extinction event of fire when He folds and puts it away at the end of the one-thousand-year millennial reign of Jesus Christ, and our current Era will then vanish like smoke. God will then put on a new garment— He will make a new and improved Era. (Isaiah 51:6; Zephaniah 3:8; Revelation 20:9, 21:1)

The second aspect refers to God's past garments of creation that were changed with mass extinction events when He folded and put them away long ago. The abundant fossils from the Triassic, Jurassic and Cretaceous folds of Earth are evidence of these anciently folded garments of God. (Psalms 90:2-6; 104:24-32)

Geological evidence indicates that God has worn five major garments of creation, one for each of the five previous Eras, and changed each of His garments of creation with a sudden mass change or extinction event, and left behind a little evidence of their existence in the fossil embedded metamorphic clay within the folds of the earth (Job 28:14). Little is known of what God did in these prior Eras, but paleontologists speculate.

God's current creation shows His handiwork. God's past handiwork is revealed in the folds of Earth and offers hints of His past Eras to paleontologists. God folds His garment in the evening, and in the morning changes into a new garment and Era of creation by making new improved bursts of life and a matching atmosphere (heavens) to sustain them.

I think of this as God seeing how He could improve His iEarth 5 and then made our iEarth 6. But I suspect God, with perfect knowledge, had vast scientific and spiritual reasons for making multiple Eras, and that each Era needed the previous to flourish, and these past sequences of creation are all part of His grand plan. God has told us in advance that our iEarth 6 will be destroyed, and from its ashes will rise a new and much improved iEarth 7 for believers to live on with God. (Revelation 20-21)

"By faith we understand that the worlds were framed by the word of God, so that the things which are seen were not made of things which are visible." – Hebrews 11:3 (Notice it states "worlds"? This can also be translated as "ages," and both are plural!)

God knows the entire sequence of events from the beginning to the end. He created time, which consists in and through God, who dwells in eternity. God knows all of His works from eternity. (Isaiah 57:15; Acts 15:18)

Thinking Slowly and Carefully

We naturally like to "think fast" by latching our thoughts onto past conclusions to avoid "thinking slow." We understand new concepts when we engage in slow, deliberate and logical thinking.[45] Inventing miracles to explain the inconsistencies in Genesis is thinking fast by instinct and emotion, and so is the evolutionary response that it all just deterministically drifted. Thinking slowly is required to understand new concepts, especially with those of which we disagree. Is it not ignorant to answer something before we understand it?

With regard to matters requiring thought:
the less people know and understand about them,
the more positively they attempt to argue concerning them
– Galileo Galilei

The evidence of sound scientific findings and the Bible are now analyzed to see if Genesis can viably be understood as a chronological verbatim account written to the perspective of ancient humans.

Genesis describes things as being made, let and formed, or created. Realizing that there is a difference between made let or formed and created is trusting that God has a purpose in using these different words with different meanings, and that this helps clarify the Genesis account—

God created Earth's land and water *before* the first day, created and blessed new birds and sea creatures on the fifth day, and created and blessed the new modern humans on the sixth day. Create signifies a new significant origin.

But in the *first* four days, God made or let things be, let it become made, or let it come to pass. This indicates that God was modifying things that already existed, or if it *was* new then it wasn't a significant creative event. Make or made can also refer to the broad reproduction process of a species.

Genesis Chapter 1

Genesis 1:1 In the beginning God created the heavens and the earth.

God created the extremely precise physics of the interconnecting and expanding Universe of time, space and matter as a subset within His spiritual realm. Since all of the sons of God, including Lucifer, "shouted for joy" when Earth was created, God created angelic beings *before* Earth. (Genesis 1:1; Job 38:7)

Stephen Hawking, an atheist, concluded that the space time and matter of the Universe began and expanded in an instant of singularity when the laws of physics broke down,[46] although quantum physics may someday assist in understanding this scientific gap.

Isaiah 42:5 simply states that the Universe began when God spoke it into existence. Jeremiah 10:12 and Zechariah 12:1 explain the heavens are stretching. The Hebrew "natah" is translated as "stretched" but it can also be translated in the present tense, such as "You stretch out the starry curtain of the heavens." (Psalms 104:2 NLT) Curtain (Hebrew "doq") may refer to the connecting strings of the vast cosmic web.

Robert Jastrow, NASA's first chairman for the Lunar Exploration Committee, said in his book *The Enchanted Loom: Mind of the Universe*, "Every event can be explained in a rational way as the product of some previous event: every effect must have its cause: there is no First Cause.... This religious faith of the scientist is violated by the discovery that the world had a beginning under conditions in which the known laws of physics are not valid, and as a product of forces or circumstances we cannot discover. When that happens, the scientist has lost control. If he really examined the implications, he would be traumatized."

Genesis 1:2 The earth was without form, and void; and darkness was on the face of the deep. And the Spirit of God was hovering over the face of the waters.

The Hebrew word for was is "hayah," meaning a change that came to pass and became something else, such as in Genesis 19:26 when Lot's wife "became" a pillar of salt.

The Hebrew for without form is "tohuw," a desolation. The Lord says He formed Earth and made it not as a desolation (tohuw), but created Earth and formed it to be inhabited. (Isaiah 45:18)

Using a Hebrew Interlinear Bible and dictionary, "the earth was without form and void" means, "It came to pass Earth became a worthless undistinguishable ruin of desolation."

God doesn't say how long after He established Earth that it became a desolation. Paleontologists think Earth was last turned into an obscuring desolation 65 million years ago at the end of the Mesozoic Era (Triassic, Jurassic and Cretaceous Periods), when a meteorite struck our planet and caused a massive tsunami wave and a dense planetary cloud of dust and water vapor.[47] Biblically, it doesn't matter how long ago this happened, leaving the minutia of these time frames within the realm of science and human observation. The focus of Scripture is on how to obtain eternal life, and so just lays out this past sequence as a broad overview.

God says Earth and its atmosphere was shaken before humans, and it will happen once more in the future— "Once more (it is a little while) I will shake heaven and earth, the sea and dry land." (Haggai 2:6)

This future once more is told of in Revelation 16:18, "there was a great earthquake, such a mighty and great earthquake as had **not** occurred since men were on the earth."

God's new and improved iEarth 7 will then rise from the ashes of our iEarth 6, just as our Era rose from the ashes of the Mesozoic iEarth 5.

The heavens, the sun, moon, stars, and Earth's solid mass and water, have been created in verse 1, and the Spirit of God now hovers over the waters that were created ***before*** the first day.

God told Job— Who laid Earths' cornerstone when the morning stars sang together, when all the sons of God shouted for joy? Who shut the sea in with doors as it burst forth from the depths? When I made a garment of clouds to wrap Earth in thick darkness and fixed the rest and motion of the waves and said, 'You may come no farther, here your proud waves stop!' Have you commanded the morning since your days began, and caused the dawn to know its place, taking hold of Earth's end to shake out the wicked formed in a mold of clay, and withheld light? (Job 38:4-9)

This imagery of God's Spirit hovering in the dark over the face of the water is from a human perspective of standing on Earth and seeing God's Spirit working in the darkness and chaos, just as the Jamieson-Fausset-Brown Bible Commentary pointed out in 1871. It is not from a satellite perspective of observing Earth from space as a brilliant white, or murky light-brown, swirl of dense, thick cloud cover.

At this time Earth is densely shrouded in clouds of water vapor and dust, with vast planetary storms raging across ice and water. The Spirit of God modifies life in the waters to generate a new sustaining atmosphere that is required for the new life forms of His next Era. Even today most of Earth's oxygen comes from algae and marine plants. Water-based life needs light, so guess what happens next in a precise scientific fashion—

Genesis 1:3-5 Then God said, Let there be light; and there was light. And God saw the light, that it was good; and God divided the light from the darkness. God called the light **Day (solar day)**, and the darkness He called Night. So the evening and the morning were the first **day (time)**.

The interlinear Hebrew word order for Genesis 1:5 is, "called God light **day** (yowm) darkness called night **it came to pass** (hayah) evening **became** (hayah) morning **Age** (yowm) **One**."

Note that God Let there be light. He let it be made to become light because He had already created the stars and the Sun, Earth and Moon, and Earth's water and solid mass before this in Genesis 1:1. Genesis is written to a perspective of those watching the sun rise and go down. Did these ancient humans know Earth orbits the Sun? The most they could see of the planet was from high on a mountain, or far out in the ocean where Earth's curvature encircled them. So, from our perspective of standing on Earth, God lets enough sunlight (**day**) through the dense clouds to nurture atmospheric-producing life forms.

The Hebrew hayah means "come to pass" or "become."

The Symbolic Meaning of Evening and Morning

The Strongs Concordance defines the Hebrew "yowm" as— sunrise to sunset, day, days, season, process of time, as long as, old, an age, chronicles, a whole age, forever or everlasting. Genesis 4:3 says yowm is "in the process of **time**." Genesis 18:11 says yowm is "**age**." Isaiah 30:8 says yowm is "a **time** to come, forever and ever." The context of each passage must be studied to determine how long of a time it is.

Every passage of scripture with both **day** (yowm) and **night** always refers to a solar day. But, *also* in Genesis 1 yowm is associated with **evening** (Hebrew "ereb") and **morning** (Hebrew "boqer"):

Evening is the **time** of dusk and burnt sacrifices, cleansing, and the cloud of the Lord. (Ezra 9:4, Daniel 9:21; Leviticus 11:24; Numbers 9:15-22).

Morning is the **time** after sunrise when one rises to change garments and begins work, or it can refer to angels.
(Leviticus 6:9-12; Ruth 2:7, 3:14; Job 38:7)

The Song of Moses explains this morning and evening symbolism—God gave birth to Earth and the world and brought mountains forth. For God, a thousand years is like a brief **age** of the past; God sweeps mortal beings away; like plants we spring up in the **morning** to bloom and flourish, and in the **evening** we are cut down and wither. (Psalms 90:2-6)

The six **ages** in which evening became morning were as long as God wanted them to be. Insisting they were twelve or twenty-four hours, or one-thousand-years, is like playing a flute and telling God we want Him to dance to our tune. God just tells us of His sequence for these six Genesis **ages**, but He did give humans the desire and ability to try and find out how long they were. (Luke 7:32; John 11:9; Psalms 19:2)

Day is solar light shining on Earth. **Age One** is the evening becoming morning. So, the first yowm means **daytime** and the second yowm means an **age** of indeterminate length.[10,44]

Genesis 1:6-8 Then God said, "Let there be a firmament in the midst of the waters, and let it divide the waters from the waters. Thus God made the firmament, and divided the waters which were under the firmament from the waters which were above the firmament; and it was so. And God called the firmament Heaven. So the evening and the morning were the second day. (Age Two)

The firmament of Heaven is our vital atmosphere that separates the oceans from the clouds. Ancient people couldn't view satellite photos of Earth and did not know what was in Heaven, which, in the context of this verse, is the atmosphere. Earth was an intense storm planet of dense clouds and rain. Then God lets light reach water in the first age so marine algae generates an atmosphere in the second age to support new life forms in the third age. Simple, right? It seems scientists may have backwards engineered Genesis and now claim it as their own, but it was right here all along.

Genesis 1:9-13 Then God said, "Let the waters under the heavens be gathered together into one place, and let the dry land appear"; and it was so. And God called the dry land earth, and the gathering together of the waters He called Seas. And God saw that it was good. Then God said, "Let the earth bring forth grass, the herb that yields seed, and the fruit tree that yields fruit according to its kind, whose seed is in itself, on the earth"; and it was so. And the earth brought forth grass, the herb that yields seed according to its kind, and the tree that yields fruit, whose seed is in itself according to its kind. And God saw that it was good. So the evening and the morning were the third day. (Age Three)

After making an atmosphere for two ages, God let the waters be gathered so land appears and then lets plants come to life. Some of these plants may have been modified from plants of the last Era. Were insects also made to maintain and pollinate them? Maybe. Genesis gives a broad overview and doesn't explain the minutiae. God didn't create or move the land into place, He let it appear as the waters receded.

Genesis 1:14-19 Then God said, "Let there be lights in the firmament of the heavens to divide the **day** from the night; and let them be for signs and seasons, and for days and years; and let them be for lights in the firmament of the heavens to give light on the earth"; and it was so.

Then God made two great lights: the greater light to rule the day, and the lesser light to rule the night. He made the stars also. God set them in the firmament of the heavens to give light on the earth, and to rule over the **day** and over the night, and to divide the light from the darkness. And God saw that it was good. So the evening and the morning were the fourth day. (Age Four)

The sun, moon and stars were already created in Genesis 1:1, so from the perspective of a human standing on Earth, God let the solar light (day), moonlight (night), and starlight be made visible.

God confirms in Job 38 that Earth was enshrouded in dense clouds and darkness in Genesis 1:2, but now these land and water-based plants have cleansed the air and generated enough atmospheric gases to allow enough sunlight through so stars are now visible. God makes a suitable atmosphere in the first three ages to prepare for new life forms. Making Earth, as they say in *Star Trek*, into a Class M planet.

Genesis 1:20-23 Then God said, "Let the waters abound with an abundance of living creatures, and let birds fly above the earth across the face of the firmament of the heavens." So God created great sea creatures and every living thing that moves, with which the waters abounded, according to their kind, and every winged bird according to its kind. And God saw that it was good. And God blessed them, saying, "Be fruitful and multiply, and fill the waters in the seas, and let birds multiply on the earth." So the evening and the morning were the fifth day. (Age Five)

God is back to creating on the fifth day. He creates the origins of new significant living sea creatures and birds, materializing them into existence after making the atmosphere to support them. These large sea creatures and birds are significant because they are blessed by God.

Genesis 1:24-31 Then God said, "Let the earth bring forth the living creature according to its kind: cattle and creeping thing and beast of the earth, each according to its kind"; and it was so. And God made the beast of the earth according to its kind, cattle according to its kind, and everything that creeps on the earth according to its kind. And God saw that it was good.

Then God said, "Let Us make man in Our image, according to Our likeness; let them have dominion over the fish of the sea, over the birds of the air, and over the cattle, over all the earth and over every creeping thing that creeps on the earth." So God created man in His own image; in the image of God He created him; male and female He created them.

Then God blessed them, and God said to them, "Be fruitful and multiply; fill the earth and subdue it; have dominion over the fish of the sea, over the birds of the air, and over every living thing that moves on the earth."

And God said, "See, I have given you every herb that yields seed which is on the face of all the earth, and every tree whose fruit yields seed; to you it shall be for food. Also, to every beast of the earth, to every bird of the air, and to everything that creeps on the earth, in which there is life, I have given every green herb for food"; and it was so.

Then God saw everything that He had made, and indeed it was very good. So the evening and the morning were the sixth day. (Age 6)

The humans created in God's image were given tree and plant seeds to eat. They populate and dominate Earth as God instructed them to do. They may have had long lives with highly developed paralimbic system to enable superior processing of amiable emotions, goal setting, motivation and self-control, similar to that of dolphins[33] which God had created and blessed. The dominion God gave these humans over other creatures may have caused the extinctions of large megafaunal beasts, some were found with what appears to be human-inflicted injuries.[36] Starting around 150,000 BC, this human expansion slowed by 10,000 BC, setting the stage for God's making of Adam around 5500 BC. Adam and Eve's punishment of God's curse of weeds on the Mesopotamian soil has not yet occurred. (Genesis 3:16-19)

The Hebrew "behemah" is a broad term for quadruped (four-legged) animals, but is translated here as cattle. Genesis begins with herbivores, or plant eaters. The carnivores came later, which paleontology supports.

We humans tend to judge the things we don't know as unimportant and not worthy of much consideration. We have tended to assume, from our perspective, that humans are the center of God's Universe and the most important of all God's creatures. But we know little of God's plans with His other creatures, such as the Genesis 1 humans.

God's spiritual dimension and the universe are big places with things happening in them that we modern humans know little of. When we get an inkling of how big it all really is, we often retreat into our own quiet little lives and try to forget about it. This book does the opposite.

Hey, aren't you guys supposed to be amiable?

Create, Make and Let

First God created the Universe and Earth.

Then Earth was a desolated ruin sitting in darkness.

Then during the first four days, God let things become made into something else. He modified or made new things.

On the fifth day God created birds and large sea creatures, and on the sixth day God made living creatures and created humans.

Create (Hebrew bara) is God creating the origins of something significant and new. God creates new sea creatures, birds and humans, and shows their significance by blessing them.

Let or made (Hebrew hayah) is when God lets it become made. God makes something from something else.

Make or made (Hebrew asah) is to do or make in the broadest sense and widest application, like the broad reproduction process of a species.

Form (Hebrew Yatsar) is to mold clay into a form, such as a clay pot.

Knowing the difference and implications of these words is vital, but assuming they are interchangeable is fatal in understanding the Genesis creation account. For example, God first says, "Let Us make man," referring to the broad repeating process of reproduction and the making of future humanity as a whole (asah), but then later signifies the actual origin of humanity when He created (bara) the actual sixth day humans. Per Isaiah 43:1, the Lord created Jacob, the origin of the Israel nation, and formed the nation of Israel from Jacob. Also, the Lord formed Adam in Genesis 2:7, unlike the created humans of Genesis 1.

Most initial plants and animals must have been made within a short time of each other due to their vast symbiotic relationships.[49] How did these creatures come into being? It logically follows that God genetically modified some from other similar existing creatures, perhaps existing creatures gave birth to and raised these new modifications, but for some, I picture an instantaneous quantum-entanglement materialization like the teleportation beaming of *Star Trek*. If so, I would love to have seen God "energize" these first humans into existence for His new Era. Adam was made of clay and then the Lord breathed life into him. However it was done, it would have been amazing to watch. Why is it we can be entertained by *Doctor Who*, yet can find it hard to believe that God, the true Time Lord, created and made all things?

Earth's Current Creatures Are Not Evolving

A 2018 scientific study released shed some light on how long modern humans may have existed. The study, "Why should mitochondria define species?," was published in the journal ***Human Evolution*** by lead author Mark Stoeckle from The Rockefeller University in New York, and David Thaler at the University of Basel in Switzerland.[7]

After examining millions of DNA samples from 100,000 different species from all over the planet, Stoeckle and Thaler concluded that **most of our current animal species emerged nearly simultaneously within the last 200,000 years.** They found 90 percent of Earth's existing animals, including humans, all emerged from small populations between 100,000 and 200,000 years ago. They also found **people and animals separated by geographic distance have not become more genetically diverse during the last 200,000 years**!

Their study and conclusions are based on neutral mutation testing which measures slight DNA changes occurring across generations. Neutral mutation testing measures DNA "tree rings" or "barcodes" that indicate how many generations old a species is.

"This conclusion is very surprising, and I fought against it as hard as I could," said Thaler. He was surprised that species have such very clear genetic boundaries, and that it is difficult to find intermediates (missing links), saying, "**If individuals are stars, then species are galaxies.** They are compact clusters in the vastness of empty sequence space." They found, similar to phone numbers, that each species' DNA sequence goes specifically to that species and none other.

Thaler also commented that the absence of intermediates is something Darwin struggled to understand— "Experts have interpreted low genetic variation among living humans as a result of our recent expansion from a small population in which a sequence from one mother became the ancestor for all modern human mitochondrial sequences."

This supports the Genesis statement that species were created or made by God to genetically continue "according to their kind," and are not genetically drifting randomly into new species. (Genesis 1:25)

This ironically supports the findings of Georges Cuvier, the founding father of modern paleontology, who argued in his Great Debate against evolutionary theory in AD 1830 that our current species are not evolving,[55] some twenty years before Charles Darwin published *The Origin of Species.*

The Eight Genesis Events of our Era

The Bible and science both state life first began in the ocean (Job 38:8 says the sea burst forth and issued from the womb), then land appears and plants come next, then living creatures, and humans last.

The eight main Genesis time events in the making and creating of our current sixth Era occur in this scientific sequence—

Light ⇒ Atmosphere ⇒ Land ⇒ Plants ⇒ Atmospheric Change ⇒ Sea Creatures and Birds ⇒ Mammals ⇒ Humans

The first five events prepare for the dependent living creatures.[27]

There is only a mere 1 in 40,320 chance of Genesis having these eight events in this order, which modern science just so happens to match.

This powerful evidence gives pause for consideration.

God broadly refers to our entire current sixth Era as being made and not created— "The Lord made the heavens and the earth, the sea, and all that is in them in six days, and rested the seventh day." (Exodus 20:11)

Does God regard our current Era as a modification of past Eras? Does our Era begin after the obliteration of most life forms at the end of the Mesozoic Era? It is very interesting that the six ages in Genesis 1 explains not only the sequence of events to produce the life forms and animals and humans for our current Era as we now know it, but it also simultaneously gives a broad overview for all six major Eras. So, not only does the six days of Genesis give a broad overview of the exact sequence for how our **current sixth Era** was made, but also at the same time gives a broad overview of how the sequence of all **six major Eras** were made, and does so in the same order and sequence as science says that they occurred! This dual application stems from "yowm" meaning an unspecified length of time, such as an age.

The exception paleontologist may find is with God's creation of birds on the fifth day, since many scientists have generally believed reptilian dinosaurs were the prominent land-life forms of the fifth Era. But Genesis indicates birds were the significant land-life forms of the fifth Era.

If you have all this evidence and proof positive that God exists, you don't need faith. I think He kind of designed it so that we'd never be able to prove His existence. And I think that's really cool.
– Dr. Mary Schweitzer

The Six Major Eras

God's wisdom is often multifaceted, and His truth multidimensional with many applications. This seems to be true of the six **ages** of Genesis, which explains both the broad sequence of events for both our current sixth Era, and a concise synopsis for all six of Earth's major Eras—

Era 1 Proterozoic Era/Eon (Age)
God's Spirit hovers over the ocean waters and initializes atmospheric producing life forms, and allows nourishing light to penetrate though the planet's blackening storm soup.

Era 2 Paleozoic Era (Cambrian, Ordovician, Silurian)
Atmospheric producing water-based plant life is made with sea life to maintain and keep them in balance, and an atmosphere is produced under Earth's ultra-dense clouds to prepare for the next Era's life forms.

Era 3 Paleozoic Era (Devonian, Carboniferous)
Dry land appears and advanced land plants are made to thrive in an oxygen rich atmosphere, along with the insects and small creatures needed to maintain and balance these plant populations.

Era 4 The Permian Marine Extinction Gap
The atmosphere clarifies and allows more sunlight and new types of life to be sustained. This important marine and plant life transition prepares an ecosystem and atmosphere to support the creation of large sea and land life on Days 5 and 6. This vital Era is sometimes overlooked.

Era 5 Mesozoic Era (Triassic, Jurassic, and Cretaceous)
The creation and blessing of larger sea creatures and birds (dinosaurs).

Era 6 Cenozoic Era (Tertiary and current Quaternary Period)
The making of our more developed sea and land mammals, reptiles, and birds, and finally the creation and blessing of humans.[27]

The ancient Hebrew Genesis text lays out a broad overview of the major ages and creatures in their order of occurrence long before scientists speculated on their sequence. Just another 1 in 40,320 coincidence, or is this more powerful evidence?

Genesis 1 is a dual simultaneous elucidation for explaining both—

1) The six-stage development of our current Cenozoic Era.
2) The six stages and generations for all six major Eras.

God, in His manifold wisdom, intended "yowm" to mean sunrise to sunset, day, days, season, as long as, process of time, old, an age, a whole age, chronicles, forever or everlasting. Knowing "yowm" has these multiple meanings helps some of us understand this dual application for the six-event sequence of Genesis. Each **time** or **age** was millions or thousands or hundreds of years, or several hours, however long God needed and wanted it to be. At the eventide of the five previous Eras, each Era withered and was consumed in a sacrificial mass extinction, and each old Era became the foundation for the next new Era to flourish on. With the dawn of each new Era, God folded His old garment of creation and put on a new garment. The fossilized layers are evidence of God's five ancient garments of creation. (Leviticus 6:9-12; Psalms 90:6)

In the six-day making of our current Era, eventide allows time for God's work to cleanly develop as the evening progresses into morning in preparation for the next age to make or create new life forms.

There are three sets of six:

1. ~Six Eras including our current Era. The first five Eras ended with mass extinction/transitional change events, as will our sixth Era.
2. ~Six time events for the making of our current human Era.
3. ~Six thousand years for our current human period.

In God's manifold wisdom, Psalms 102:25-26 clarifies that God changed Earth *and* its atmosphere (heavens) for new Eras— "Of old You laid the foundation of the <u>earth</u>, and the <u>heavens</u> are the work of Your hands. They will perish, but You will endure; yes, they will all grow old like a garment; like a cloak You will change them, and they will be changed." God made the oxygenated Carboniferous Era with its dragonflies of twenty-eight-inch wingspans, the giant redwoods and dinosaurs of the Jurassic Era, and our sixth Era. God's new and much improved iEarth 7 is yet to come. (Revelation 20-21)

Truth is the offspring of silence and meditation.
I keep the subject constantly before me and wait
'til the first dawning opens slowly,
by little and little,
into a full and clear light
- Isaac Newton

The Big Stretch

Since ancient times Christians believed the Universe began when God created it. But for hundreds of years astronomers observed the orderly rotation of stars around the center of our galaxy and logically concluded that the Universe was eternal, or static, and therefore had no beginning. They assumed the galaxies were simply stars and our Milky Way Galaxy was the entire Universe.

Albert Einstein also believed the Universe had no beginning. Einstein had rejected the creation story of Genesis when he was twelve and concluded the Bible was a primitive collection of childish legends. So, when he applied his general theory of relativity to the Universe in 1915 and discovered that the Universe is either expanding or contracting, he simply altered his equation and put in a cosmological constant factor to hold his eternal Universe steady. But, a few years later Alexander Friedman and George Lemaitre developed models from Einstein's theory showing the Universe actually *is* expanding. Astronomer Edwin Hubble soon confirmed their models were correct, and that the Universe truly extends well beyond our own Milky Way galaxy! Hubble showed how the red light shift in some stars proved that they are actually rapidly retreating galaxies, and Universe is flying apart at vast velocities! So Einstein visited Hubble at the Mount Wilson Observatory in California, and then concurred with Hubble that the Universe is indeed expanding.[52] (Isaiah 42:5; Jeremiah 10:12; Zechariah 12:1; Psalms 104:2)

Einstein later wrote of how he desired "to know how God created this world. I am not interested in this or that phenomenon, in the spectrum of this or that element. I want to know His thought; the rest are details." Einstein believed in a god that is revealed in the lawful harmony of the world, but not in a personal God.

Fred Hoyle, the brilliant astronomer who derisively named this expanding Universe theory "The Big Bang" fervently opposed it his entire life. He explained, "The reason why scientists like the "big bang" is because they are overshadowed by the Book of Genesis. It is deep within the psyche of most scientists to believe in the first page of Genesis." Hoyle also wrote, "The big bang theory requires a recent origin of the Universe that openly invites the concept of creation."

George Gamow predicted that *if* the Big Bang did occur, then the cosmic background temperature of the Universe should be a few degrees above absolute zero. This was confirmed in 1965 in part by Nobel Prize winner Arno Penzias, who said, "The best data we have are exactly what I would have predicted had I nothing to go on but the first five books of Moses, the Psalms, and the Bible as a whole."

George Smoot, who later won a Nobel Prize for *his* work on cosmic background radiation, wrote, "The big bang, the most cataclysmic event we can imagine, **on closer inspection appears finely orchestrated**."

This confirmation of the Big Bang theory was a shock to scientists because it proves time began when the Universe emerged from a very fine point when the space-time curvature, temperature, density and pressure was infinitely small. This is referred to as a singularity.

The Hubble Law is one of the great discoveries in science;
it is one of the main supports of the scientific story of Genesis
- Robert Jastrow, of NASA

Per *hawking.org*, Stephen Hawking, an atheist, said, "All the evidence seems to indicate that the Universe has not existed forever, but that it had a beginning about 15 billion years ago. This is probably the most remarkable discovery of modern cosmology. Yet it is now taken for granted." Stephen Hawking said many scientists are unhappy with the Universe having a beginning because it implies physics broke down and "one would have to invoke an outside agency, which for convenience, one can call God, to determine how the Universe began," so scientists "therefore advanced theories in which the Universe was expanding at the present time, but didn't have a beginning." Hawking explained, "The real reason we are seeking a complete theory is that we want to understand the Universe and feel we are not just the victims of dark and mysterious forces. If we understand the Universe then we control it, in a sense. The standard model is clearly unsatisfactory in this respect."

For the scientist who has lived
by his faith in the power of reason,
the story ends like a bad dream.
He has scaled the mountain of ignorance;
he is about to conquer the highest peak;
as he pulls himself over the final rock,
he is greeted by a band of theologians
who have been sitting there for centuries
- Robert Jastrow, of NASA

Birds v. Reptiles

Although evolutionary theory logically dictates that flying birds were not the first life forms evolving from ocean creatures, evolutionists are unsure at what point after this that the timing and development of birds took place, in part because scientists haven't found evidence of the widespread existence of birds in the Triassic, Jurassic, and Cretaceous Periods.

Some birds are known to exist then, and a few birds may have existed in the Triassic period, but the problem with finding bird fossils is that their tiny bones make their fossils very rare and difficult to find, so this issue may never be adequately resolved.[50]

But, if a dual application is made for the six days of Genesis to also dually describe all six major Eras, then a conflict arises for the fifth Era—Genesis states birds were the significant land creatures during the fifth Mesozoic Era, but scientists have traditionally thought reptilian dinosaurs ruled the Triassic, Jurassic, and Cretaceous Periods (the Mesozoic Era).

Birds being "created" and "blessed" on day five indicates that birds were the significant land creature event of the fifth day (Mesozoic Era), just as humans were the significant land creature event of the sixth day (our current sixth Era). Genesis using "create" indicates the origin and importance of these birds, as opposed to the mammals on day six that were just "made," indicating that mammals played a minor role, and also perhaps that it wasn't their origin but rather that they were modifications from prior mammals of the Mesozoic Era.

This bird conflict becomes quite interesting when matching the fifth day of Genesis to the Triassic, Jurassic and Cretaceous Periods of the Mesozoic Era. Some scientists speculate that dinosaurs and birds are closely related.[50,51]

Mary Schweitzer, the respected paleontologist who first discovered soft tissue in a T-rex bone, describes herself as "a complete and total Christian." Dr. Schweitzer said, "If you have all this evidence and proof positive that God exists, you don't need faith. I think He kind of designed it so that we'd never be able to prove His existence. And I think that's really cool." Her astonishing discovery of soft tissue in T-rex bones led to another discovery, that the T-rex may have had red blood cells similar to birds, such as ostriches. Dr. Schweitzer is a fundamental Christian who does not believe in young earth creation theories.[51]

Finely Adjusted Physics

What rational reasons are there for the precise fine-tuning of the fundamental physical laws of the Universe to even exist and function, such as the "knife-edge" rate of its expansion and the force of gravity, electromagnetic forces, the ratio of neutrons and protons, strength of nuclear force, cosmic parameters, the carbon and oxygen process ratio, and among many others, the hydrogen to helium process?[53]

Although scientists all agree that physical laws are fine-tuned, for a few scientists it indicates a Creator might be behind them—

Fred Hoyle, the astronomist who discovered nucleosynthesis in stars and vehemently opposed the Big Bang and creation– "A common sense interpretation of the facts suggests that a super intellect has monkeyed with physics, as well as with chemistry and biology, and that there are no blind forces worth speaking about in nature."

Stephen Hawking, renowned astrophysicist, atheist and author of *A Brief History of Time: From the Big Bang to Black Holes*– "The laws of science, as we know them at present, contain many fundamental numbers... The remarkable fact is that the values of these numbers seem to have been finely adjusted to make possible the development of life."

John Aloysius O'Keefe III, expert in planetary science and astrogeology with NASA– "If the Universe had not been made with the most exacting precision we could never have come into existence."

Wernher von Braun, pioneer rocket engineer– "I find it as difficult to understand a scientist who does not acknowledge the presence of a superior rationality behind the existence of the Universe as it is to comprehend a theologian who would deny the advances of science."

Vera Kistiakowsky, MIT physicist– "The exquisite order displayed by our scientific understanding of the physical world calls for the divine."

Alan Sandage, winner of the Crawford prize in astronomy– "I find it quite improbable that such order came out of chaos. There has to be some organizing principle. God to me is a mystery but is the explanation for the miracle of existence, why there is something instead of nothing."

Antony Flew, professor and author of *There Is a God: How the World's Most Notorious Atheist Changed His Mind*– "It now seems to me that the findings of more than fifty years of DNA research have provided materials for a new and enormously powerful argument to design."

Arthur L. Schawlow, Professor of Physics at Stanford University, 1981 Nobel Prize in physics– "It seems to me that when confronted with the marvels of life and the Universe, one must ask why and not just how. The only possible answers are religious. . . I find a need for God in the Universe and in my own life."

Ed Harrison, cosmologist and author of *Masks of the Universe*– "The fine tuning of the Universe provides prima facie evidence of deistic design. Take your choice: blind chance that requires multitudes of universes or design that requires only one.... Many scientists, when they admit their views, incline toward the teleological or design argument."

Frank Tipler, author of *The Physics of Christianity* and *The Physics of Immortality*– "When I began my career as a cosmologist some twenty years ago, I was a convinced atheist. I never in my wildest dreams imagined that one day I would be writing a book purporting to show that the central claims of Judeo-Christian theology are in fact true, that these claims are straightforward deductions of the laws of physics as we now understand them. I have been forced into these conclusions by the inexorable logic of my own special branch of physics."

Henry Schaefer III, director of the Center for Computational Quantum Chemistry at the University of Georgia, and lecturer of *Scientists and Their Gods*– "The significance and joy in my science comes in those occasional moments of discovering something new and saying to myself, 'So that's how God did it.' My goal is to understand a little corner of God's plan."[54]

Scientists stay clear of "stop thinking" solutions such as "God made it," since they desire to think and discover, just as God intended.

Scientists who think in terms of how and why God did it *are* true to their profession, don't waste time pursuing false solutions, and are truly open minded to all possibilities. However, when it is discovered *how* God did it, *some* scientists simply assume it happened all on its own!

The more time and devotion one spends
in the worship of false gods,
the less he is able to spend
in that of the True One
- Isaac Newton

How and Why

The evolutionary paradigm is— *How* did random genetic processes interact with deterministic processes in the natural selection of organic life, and *how* do all living creatures share universal common descent? (Is not calling it a "*descent*" a biased assumption it evolved? Would not calling it a "*relationship*" be scientifically precise?)

The theologian view is— *Why* did God determine genetic processes with universally common *authorship*?

Renowned quantitative evolutionary paleontologist David Raup concluded in his book *Extinction, Bad Genes or Bad Luck*, that extinctions resulted from either bad genes or bad luck— "At no point in the fossil record can we look at a particular event and say, 'Of course, it had to happen that way.'" He also stated, "The disturbing reality is that for none of the thousands of well-documented extinctions in the geologic past do we have a solid explanation of why the extinction occurred."[22]

The broader question is not *how* five meteorite impacts caused five mass extinctions,[22] but *why* did God use these meteorites, or some other means, to cause these mass extinctions? In the future God has four angels stop Earth's four winds just before Earth is scorched and some objects from space begin falling to Earth.[3,4,5] (Revelation 7:1-2; 8:1-13)

Finding out *how* it was done doesn't prove or disprove God! Considering *why* God did it is the deeper question. For example, the highly sophisticated 500-million-year-old trilobite doublet-eye-lens system is comparable to what a talented optical engineer might now design and was not primitive at all.[22] Both atheists and creationists leap forward to debate *how* this could be. But, *why* did God make trilobites? What if paleontologists evaluated the quantitative data from past Eras to consider *why* they were designed by God and what their function was in that world? What roles might be discovered where atheists, in eliminating God, missed in their focus on *how*? What would be left of our intelligence if humans had existed millions of years ago? Even the trace elements left in the soil from the concrete and corroded metals of our cities would be disturbed by geological forces. We really don't know what, if any, intelligent life God has created in the past. God only given us clues.

My whole life has been guided by the principle of Plato's Socrates: Follow the evidence, wherever it leads – Antony Flew

The God's Garments View of Creation

After creating the Universe and suspending Earth in space, God made six amazing Eras— Imagine rising in the morning and folding your old robe, and then putting on a new robe, for five days in a row. In this way God made the five major Eras before ours, folding each old Era on top of the last. (Leviticus 6:9-12; Job 26:7)

So, it came to pass in the process of time that the eventide of the fifth Era ended, and as dawn began God again took up His artistic work to make, let things be and create, to renew the face of Earth for His sixth Era. (Psalms 104:29-30 paraphrased and transliterated)

> **Genesis 2:1-4** Thus the heavens and the earth, and all the host of them, were finished. And on the seventh day God ended His work which He had done, and He rested on the seventh day from all His work which He had done. Then God blessed the seventh day (age) and sanctified it, because in it He rested from all His work which God had created and made. This is the history of the heavens and the earth when they were created, in the day (yowm) that the Lord God made the earth and the heavens,

God doesn't get weary, so His rest in this seventh age was to allow His sixth Era to flourish. (Isaiah 40:28)

Does in the day (yowm) refer to that time when God made Earth and the heavens in six ages?

> The Hebrew "towledah" is often translated as "generations" instead of history since "towledah" is used to denote a genealogy— a sequence of births, or a line of descent. Genesis 1 explains the six "generations" or "sequence of births" or "line of descent" of Earth's Eras and matching atmospheres. (Hebrews 11:3)

The future Great Tribulation ends soon after space bodies hit Earth, and a massive earthquake levels the mountains and islands. But the sixth planet-wide mass extinction event occurs one thousand years later after the Millennial Reign of Christ, when Earth is renovated by fire and a new and improved seventh Era takes its place. There is more on this in the *God's Master Plan* chapter, "The Spirit and Super Body."

Why is that comma there after heavens at the end of Genesis 2:4?!

4

Genesis 2

Why does the summary of God's seven ages end with a comma after "heavens" at the end of verse 4? Here is the text again:

Genesis 2:1-7 Thus the heavens and the earth, and all the host of
them, were finished. 2 And on the seventh day God ended His work
which He had done, and He rested on the seventh day from all His work
which He had done. 3 Then God blessed the seventh day and sanctified
it, because in it He rested from all His work which God had created and
made. 4 This is the history of the heavens and the earth when they were
created, in the day that the Lord God made the earth and the heavens,
5 before any plant of the field was in the earth and before any herb
of the field had grown. For the Lord God had not caused it to rain on
the earth, and there was no man to till the ground; 6 but a mist went up
from the earth and watered the whole face of the ground. 7 And the Lord
God formed man of the dust of the ground, and breathed into his nostrils
the breath of life; and man became a living being.

The original Hebrew did not have chapter and verse notations or punctuation. These were added in later by translators who used their own judgement.[56] But the original scripture was authored by men of God who spoke as they were moved by the Holy Spirit. (2 Peter 1:20-21)

The chapter and verse and comma placement in Genesis 2 sure seems to desperately be trying to connect Genesis 2:3 with 2:4 and 2:5!

Why doesn't Genesis chapter 2 begin with what is now Genesis 2:4? Who decided where the Bible's chapters and verses should be?

Paratext

Paratext is notations such as chapters and verses that were added later on to the original text. The Torah, the first five books of the Bible, did not originally have paratext, but was divided later into 154 sections to ease reading during long religious ceremonies that lasted for days.[56] Having chapter and verse numbers are useful reference guides, but can they be relied upon for interpretation? Who inserted them, and when?

The AD 1008 Leningrad Codex has paragraph indicators after each of the seven Genesis 1 events, and ends with what is now Genesis 2:3.

Notation in the Septuagint Alexandrinus Codex of AD 440 indicates that Genesis 2 should begin with what is now Genesis 2:8— The Greek Septuagint LXX was done for the Royal Library of Alexandria in AD 200, the ancient world center of Greek thought and culture. Septuagint means "seventy" in Latin, and was named because seventy-two Jews translated it from Hebrew. Apostles quoted from the Septuagint. It was highly regarded and used by the early church until about AD 500. Augustine spoke of the Septuagint in his *City of God* book of AD 426, "For the same Spirit who was in the prophets when they spoke these things was also in the seventy men when they translated them."

However, our current chapter placement for Genesis 2 is based on paratext from Latin Vulgate editions— Stephen Langton, an English Cardinal and Archbishop of Canterbury, is traditionally credited with making this decision in about AD 1205.

So, chapter and verse placement can't be depended on to interpret the Bible, unless one has confidence in the AD 1205 judgment of Stephen Langton, and perhaps other unknown scholars along the way who made notations in their copy of the Latin Vulgate or Septuagint.[57,58,59,60,61,62,63] Langton's biased chapter placement of Genesis 2 does seem to connect Adam and Eve to the six days of creation, but this connection did not exist in the original Hebrew text! Where would the paratext have been inserted if translators thought Adam was made *after* Genesis 1?

There have been very few translation errors in the past. The King James 1611 version, for example, only has about 150 printing errors, spelling standardizations or minor phrase changes, but less than half of these are significant. 75 changes may seem like a lot, but this is just one hundredth of a percent of the 791,328 original words, a very small margin of error indeed.[64]

A Proposed Translation

This Interlinear King James Version of Genesis 2:5-7 is in the same word order as the Hebrew —

And every plant of the field before it was in the earth and every herb of the field before it grew for not had caused it to rain the Lord God upon the earth and a man there was not to till the ground but a mist there went up from the earth and watered the whole face of the ground. **And** formed the Lord God this man of the dust of the ground and breathed into his nostrils the breath of life became and man a soul living.

The New American Standard Bible (and Interlinear) of Genesis 2:5-7—

Now no shrub of the field was yet in the earth, and no plant of the field had yet sprouted, for the LORD God had not sent rain upon the earth, and there was no man to cultivate the ground. But a mist used to rise from the earth and water the whole surface of the ground. Then the LORD God formed man of dust from the ground, and breathed into his nostrils the breath of life; and man became a living being.

My proposed translation of Genesis 2:5-7 is consistent with the NASB Interlinear and Mechon Mamre Hebrew translations—

And every plant of the field was not yet in the land, and every herb of the field had not yet sprouted, because Yahweh God had not yet caused rain on the land, and a man there was not to cultivate the soil. Mist rose from the ground and watered the entire soil surface. Then the Lord God formed a clay man from the soil, breathed life into him, and this man became a living soul.

If this passage is referring back to the sixth day, then that should be clear from the Hebrew text, not from strategic chapter and verse placement! The portrayal of Adam being created on the sixth day reflects the translators' beliefs of that time. Just as they believed the Sun orbited Earth, they may have also believed Adam's account was elaborating on the sixth day and translated accordingly. However, these chapter and verse and punctuation choices have caused confusion, disputes, books, and ostentatious museums.

Field Plants

Genesis 2:5-6 says no agricultural field plants and herbs, or vegetables, were in the land, and there was no farmer to cultivate them. God had already made plants and trees in Genesis 1 during the third age. God gave the Genesis 1 humans plant and tree seeds to eat, and enabled them to spread out and find food. They eventually filled and subdued the planet as God had told them to do. Starting out as gatherers of plants, they may have later eaten animals since God gave them dominion over all animals, although God's stated diet for them was vegetarian.

After the Genesis 1 humans had populated Earth, the Ice Age pushed them towards the equator. Then, a subsequent climate stability of warming and drying about ten thousand years ago allowed for population expansion back into the northern and southern climates.[66]

But these expanding populations were having a difficult time finding enough seeds to gather ten thousand years ago, due to areas of vegetation being shrunk by this dryer climate.[37,38] The dry Mesopotamia plain was lushly fed by melting alpine snowmelt and glacier ice from the mountains far above. The changing rivers and ocean tides of those times caused a dire need for methodical canal systems to master the irrigation of farming small areas.[37] Modern grains, fruit trees and farm animals were needed to facilitate dense farming to assist these growing populations, and God notes there was no cultivator to till the ground. It *doesn't* state "there was no man," but "there was no man to cultivate."

Modern Man

God "forming" a man from clay with the ability and inclination to cultivate indicates he was a modification of the "created" Genesis 1 humans, but otherwise not much different. There *were* humans, but they had little inclination to cultivate the ground, design complex irrigation canals, and a governing system to oversee it. So, God began that in Mesopotamia with the making of Adam. This also gives a fascinating insight on how God may have created other humans and creatures, of making them out of clay and breathing life into them.

God breathes life into this man and then offers the choice to obey, to sin or not to sin. Why? Maybe it's because God wants our relationship with Him to be based on choosing to be in harmony with Him, and not on based on mere compliance.

Modern Fruit Trees

> **Genesis 2:8-9** The Lord God planted a garden eastward in Eden, and there He put the man whom He had formed. And out of the ground the Lord God made every tree grow that is pleasant to the sight and good for food. The tree of life was also in the midst of the garden, and the tree of the knowledge of good and evil.

Long ago in the Mesopotamian plain before Gilgamesh and the Code of Hammurabi of Babylon emerged humanities first modern civilization, and it was there that God plants these fruit trees. He did not create them, indicating modern fruit trees were modified from existing trees. Tree seed and plant food (Hebrew "**zera**") was provided to the humans of the sixth day, but field plants and tree fruit-flesh food (Hebrew "**maakal**") are now provided for Adam.

Eden and The Cradle of Modern Civilization

> **Genesis 2:10-14** Now a river went out of Eden to water the garden, and from there it parted and became four riverheads. The name of the first is Pishon; it is the one which skirts the whole land of Havilah, where there is gold. And the gold of that land is good. Bdellium and the onyx stone are there. The name of the second river is Gihon; it is the one which goes around the whole land of Cush. The name of the third river is Hiddekel; it is the one which goes toward the east of Assyria. The fourth river is the Euphrates.

In the river-silted Fertile Cresent Mesopotamia plain of the ever-changing rivers of the Euphrates and Tigris, Eden was at the time a rainless paradise with cool, clear, drinkable spring-fed ponds and high-water tables fed by the melting ice and rain from the mountains above. The cradle of modern civilization began here, the origins of the wheel, writing, sophisticated agriculture canal irrigation and crop rotations of wheat, barley, lentils, peas, flax, sheep, cows, pigs and goats, astronomy, 360 degree circles and 60 minute hours, metropolitan governments with civil service, population census, taxes, laws with the presumption of innocence, currency, trade networks and corporations, contracts, high interest loans, customer complaints, mass production of pots, sailboats, chariots, metallurgy, arched bricks, urban planning, dentistry, surgery, herbalist, maps, gods, temples, priests and rituals, beer, and Judaism.

When English archaeologist William Loftus discovered the ancient city of Uruk in 1849, the English sophisticates were bewildered by the ancient artifacts and cuneiform clay writing tablets from their excavation sites in Mesopotamia. The Fosters point out in their book *Civilizations of Ancient Iraq* that European scholars had regarded ancient Mesopotamians as simple nomadic herdsmen in ancient Iraq through the lens of the Bible. Abraham was from Ur, and Uruk was Nimrod's city of Erech. But as more clay tablets were translated, the history of Uruk and Ur rose from the ruins. They were Earth's first modern cities that had existed for some 4,000 years! English scholars first argued that other cultures had influenced these cities, but with more evidence it became clear that Uruk predates Egyptian and oriental cultures, and it was Uruk and Ur who had introduced writing and other elements of modern civilization to Egypt.

In three hundred years Uruk went from hunter gathers to a sophisticated city, and with no outside immigration! The Fosters write, "It is one of the most fascinating riddles of antiquity." It was hard to believe, as if indigenous people of AD 1600 had turned Manhattan into a modern city by AD 1900 with no outside help (p.16)!

Kriwaczek explains in his book *Babylon* how Mesopotamian knowledge and culture were channeled to the Assyrian, Babylonian, Perian, Hellenistic and Roman Empires, was the bridge over which Greece used to pass over from their archaic to classical era, and was later inherited by Europeans (p.209). The Mesopotamians explained that it was their god Lord Enki who gives rulers intelligence and knowledge and opens the door of understanding to bring abundance. He taught canal building and how to plan temple foundations in the right place. He is not the ruler of the gods, but a wise counselor and elder brother, the Lord of the Assembly who taught them how to fashion artisans, and it was he who divided the speech of mankind that had previously been one (p.30).

When gifted scholar George Smith first translated the Epic of Gilgamesh, he was so excited he ran around the room and began taking his clothes off— Here was an account of the ancient world being destroyed by a great flood, and quite similar to Noah's! These ancient clay tablets from the Babylonian library in Ninevah told of Gilgamesh's quest for eternal life and find the long-lived sole survivor of this flood to find out why the gods had granted Uta-napishtim immortality, who explained to Gilgamesh that the Flood was unique and the gods would not do that again.

Then more tablets showed King Gilgamesh of Uruk was likely real.

Then the debate ensued of "which came first, the chicken or the egg".

The egg argument is that the Epic of Gilgamesh was written 1,000 years afterwards on 3,600-year-old **clay** tablets, but still predates the writing of Genesis, so the Jewish account must have been derived from ancient myths. This argument still challenges Christianity, Islam, and Judaism.

The chicken argument is that Terah was 28 years old when Noah died, leaving about four people to pass on Noah's account to Moses. Israel's concise account was publicly recited by younger people in family Oral Traditions as their elders carefully monitored accuracy, so a high-quality realistic **living** account lived on, but Gilgamesh's exaggerated account was retold many times by obscure paid scribes who were susceptible to exaggeration and political influence from many powerful rulers of different nations and nationalities, until it was written down a thousand years afterwards. Also, Abraham could have had Noah's account written on clay tablets from Ur. (Deuteronomy 6:20-21; 32:7; Psalm 78:2-4; Joel 1:2-3)

Good and Evil

Genesis 2:15-17 Then the Lord God took the man and put him in the garden of Eden to tend and keep it. And the Lord God commanded the man, saying, "Of every tree of the garden you may freely eat; but of the tree of the knowledge of good and evil you shall not eat, for in the day that you eat of it you shall surely die."

God does not forbid eating from the Tree of Life, and says nothing of other creatures dying if they ate from the tree of the knowledge of good and evil. Since the Genesis 1 humans had no knowledge of good and evil, they have may have been similar to dolphins hanging out in the ocean and generally getting along and having a great time without sin, and who are, arguably, more intelligent than humans in some important aspects; their highly-developed neocortex and paralimbic system allows for enhanced processing of emotional self-control, goal setting, motivation, and judgment.[33](Acts 14:16; Revelation 14:4-5)

Having no knowledge of good and evil may seem like a social utopia, but since we *do* know good and evil, then having rules with consequences is vital. God's Commandments, which means "divine prescriptions," or "some things to fix the problem," are embedded within foundational philosophy for modern societies.

Having God's divine remedies, to respect God and taking time to rest, to respect parents, to be content, to know there are consequences for doing wrong regardless of who sees us, and making it wrong to kill, steal, lie as a witness, have sexual desires outside of a marriage, or long for what our neighbor has, these are all great morals for our neighbors and friends and parents and kids to live by.

It also makes sense, with my limited understanding, of why God may have allowed the knowledge of good and evil into the human race—Humans with great social skills who are thinly spread across the planet to seek food don't have much of a need for rules. But it would be chaotic to resolve conflicts in a densely populated area without standardized rules. Also, humans who know good and evil can relate more to God. But with rules from God also comes consequences for disobedience.

Modern Farm Animals

Genesis 2:18-20 And the Lord God said, "It is not good that man should be alone; I will make him a helper comparable to him." Out of the ground the Lord God formed every beast of the field and every bird of the air, and brought them to Adam to see what he would call them. And whatever Adam called each living creature, that was its name. So Adam gave names to all cattle, to the birds of the air, and to every beast of the field. But for Adam there was not found a helper comparable to him.

The Hebrew "**behemah**," which means four-legged beasts, were made on the sixth day in Genesis 1, but the Hebrew "**chayat**," which means flesh beasts for the field, are now made for Adam.

God created the humans of Genesis 1 *after* making the four-legged animals on the sixth day, but formed Adam *before* making these flesh beasts of the field in Genesis 2.

Tree and plant seeds were given as food to the Genesis 1 humans, but cultivating field plants, fruit tree flesh food, and field flesh beasts are now provided to Adam, when modern agriculture of plants and animals began.

Modern Woman

Is God noting that no animals were suitable as a helper comparable to Adam? God brought these animals for Adam to name, not to see if they could be his mate!

The phrase comparable can be translated as in the presence of or counterpart or corresponding. Consider this alternative translation, with the changes underlined—

> The Lord God said, "It is not good that man should be alone. I will make for Adam help in his presence." Out of the ground the Lord God formed every beast of the field and every bird of the air, and brought them to Adam to see what he would call them. And whatever Adam called each living creature, that was its name. So Adam gave names to all cattle, to the birds of the air, and to every beast of the field.
>
> Now for Adam there was not found a corresponding help for him.

But was added by translators to make it seem as if none of the animals measured up as a helper. "Now for Adam" is more consistent.

God forms these farm animal helpers right out of the ground in Adam's presence. But did God, infinitely wise, after making Adam with reproductive abilities and the ability to love, *really* think farm animals might work as a counterpart help for Adam?! Imagine Adam giving the thumbs up for a cow or goat! Not found is obviously referring to the Genesis 1 women. There *were* women, but none were *found* to be a suitable counterpart help for Adam.

A counterpart help is considered of small worth in barbaric societies and large corporations, but being a helper is God's highest calling. Jesus came to help humanity. God is our helper. (Psalms 46:1, 54:4)

> **Genesis 2:21-25** And the Lord God caused a deep sleep to fall on Adam, and he slept; and He took one of his ribs, and closed up the flesh in its place. Then the rib which the Lord God had taken from man He made into a woman, and He brought her to the man. And Adam said: "This is now bone of my bones and flesh of my flesh; she shall be called Woman, because she was taken out of Man."
>
> Therefore a man shall leave his father and mother and be joined to his wife, and they shall become one flesh. And they were both naked, the man and his wife, and were not ashamed.

Therefore is translated from two Hebrew words, "'al" and "ken," which have connotations of referring to something beyond this time.

A fair alternate translation is: "As in like manner, a man shall leave his father and mother and be joined to his wife, and they shall become one flesh." Since translators thought Eve was created on the sixth day, translating it As in like manner wouldn't make sense, so Therefore was used to strongly connect this to Eve, but an examination of the evidence in retrospect indicates that Adam and Eve's descendants will *also* form their own marriage units—

Jesus explained marriage began in Genesis 1 when He said, "Have you not read that He who created them at the beginning 'made them male and female,' also said, 'For this reason a man shall leave his father and mother and be joined to his wife, and the two shall become one flesh'?" (Matthew 19:4-5 and Mark 10:5-9)

The Greek "ktizo" means created and directly references the sixth age in Genesis 1 when God created them as male and female.

The Greek "kai" is fairly translated here as also.

Eve was "banah," or built, from Adam's genetics, which is similar to the "yatsar" forming of Adam. Her female descendants are the critical key to modern civilization's marriage and family units, communities, homes, flush toilets, décor, cleanliness, civility, peace and security, and men have been desiring and pursuing them ever since, wanting them to be happy and complying with their requests.

You have ravished my heart with one look of your eyes
With one link of your necklace - Solomon

Modern Human Origin

Since God doesn't get weary per Isaiah 40:28, God's seventh day of rest to allow the Pleistocene epoch to flourish likely began before the ice age from 35,000 to 10,000 BC, and the dry spell after it from 10,000 to 6,000 BC.[37] So, when Adam was formed, the large megafaunal species of the Pleistocene epoch, such as saber tooth type tigers, had died out in the Quaternary extinction of 10,000 BC, believed to have been caused by the expansion of the human race throughout the planet.[36] Domestic cows and sheep would not have coexisted that well with megafaunal species!

God overlooked the ignorance of Genesis 1 humans. (Acts 17:30)

How long ago were the Genesis 1 humans created by God?

It's not biblically documented, but science tries to date them:

1) Neutral mutation DNA testings place modern human origin from 100,000 to 200,000 years ago.[7]
2) Ice core samples taken from the climatically stable south pole in Vostok measuring the dust and bubbles of past summers,[65] similar to annual tree rings, place our current period at 160,000 years (Some compare Vostok ice to Greenland ice, but this is like comparing arid desert rainfall to tropical forest rainfall!).
3) Radioactive testings of modern human remains date them as old as 140 thousand years, but these datings vary widely.[8]

The average of these is 150 thousand years ago. Aging by fundamental evolutionists could be overstated since their theory requires many zeros in the years in order to make evolution a more probable theory.

Adam and Modern Agriculture Origin

Adam and Eve may have lived thousands of years before their banishment from the garden of Eden and the Tree of Life fruit, which was guarded after they were expelled, otherwise, God said, they might eat from it and live forever. They began to reproduce after losing their option to eat from the Tree of Life. Did God intended for Adam to be a naked farming consultant to teach the local naked Genesis 1 humans how to cultivate and build, and to supply the grains and animals for their domestic farming needs, so they could eventually terraform Earth into an Eden? If so, then any resulting children's Bible story books will need a lot of bushes for all these people to conveniently stand behind.

Genealogical dating in the Latin Vulgate places Adam's arrival around 4016 BC, which is close to Ussher's date of 4004 BC.[12] But genealogical dating per the Septuagint LXX places Adam arriving around 5500 BC. These two dating methods have been debated for some 1,600 years—

Jerome's Latin Vulgate was translated in AD 405 from the Hebrew text of his time. It's argued that when this Hebrew text was standardized by Pharisaic rabbis after the AD 70 temple fire, some Genesis genealogies were shortened by 100 years each to circumvent the timing of Jesus fulfilling the Pharisee's own "Days of the Messiah" and "Sabbatical Jubilee" prophesies which they had derived from the book of Daniel.[61]

Augustine thought the Septuagint LXX was translated in 200 BC from older and more accurate Hebrew text. It's argued that the Hellenistic Jews who translated the Septuagint LXX were influenced by Greek paganism, and so they exaggerated some genealogies by 100 years each.[61]

When does science place the origins of modern farm crops, fruits and animals?— Botanists date the origins of modern fruit trees about 3,000 to 9,000 years ago.[39] Domesticated wheat and barley origins date back to 9,000 years ago.[41] Recent DNA tests show cows originated about 10,500 years ago,[40] but are also genetically linked to ancient bovines, hence God "forming" cows. Chickens are another globally consumed farm animal traceable to about 8,000 years ago and are genetically related to junglefowl,[43] again supporting they were "formed" or "made." The same for goats and sheep, both originating about 10,000 years ago and with possibly genetic linkage to ancient animals.[42] Domestic dogs may have surfaced about 7,000 years ago.[67] Modern horses date back to about 3,000 years,[68] and genuinely perplex evolutionists.[16,69]

During Adam's time Earth's other major food crops were also being domesticated in other far reaches of the globe; Maize, squash and beans in Mesoamerica, and rice, soybeans, apricots and peaches in ancient China. But, as Roberts points out in his book *The Holocene*, only in Mesopotamia is there a strong case for a connection between their engineering marvels of hydraulic irrigation systems and a centralized political bureaucracy apparatus to manage it (p.238).

Since his youth Yuval Harari was troubled by not understanding our origins, how ancient history impacts us, and what the aim and meaning of life were. Influenced by Buddha and Darwin, and a PhD in history from Oxford, he wrote *Sapiens, A Brief History of Humankind*, explaining from a fundamentalist evolutionary perspective of how archaeological findings and dating places many ancient humans much older than Adam. Of course, he credits evolution for our origins and diversity. I credit God in creating many types of humans and other creatures in many places and times on Earth. He had a plan for them as well, but had a special plan for Adam's genealogy, which has been well documented. Who knows why God does these things? Scientific dating is generally in accord with placing Adam's origins at the beginning of, and in the same area as, the cradle of life where modern civilization began.

Coincidence? I think not.

Genesis: Abstract, Symbolic, and Literal

After Eve was deceived and Adam chose to eat from the forbidden tree of knowledge, God tells the serpent in Genesis 3, "I will put enmity between you and the woman, and between your seed and her Seed. He will bruise your head, and you will bruise His heel." This foretells of Jesus later bruising the head, or an authority, of Satan.

God told Eve, "Greatly I will increase your pain and pregnancy, in pain you will bear children, and to your husband will be your desire, and he will rule over you." With the Tree of Life gone she would now die. She already knew what pain was, but would have more pain in childbirth.

God tells Adam his punishment: "The ground is now cursed for your sake; in toil you will eat of it as long as you live. Both thorns and thistles it will grow for you, and you will eat the herb of the field. You will sweat as you eat bread until you return to the ground, for out of it you were taken. For dust you are, and to dust you will return."

Job gives insight into Adam's reaction after his choice: "If I have covered my transgressions as Adam, by hiding my iniquity in my bosom, because I feared the great multitude, and dreaded the contempt of families." - Job 31:33-34

Evidently the many people and families of Genesis 1 living in Mesopotamia were impacted and knew Adam's action had caused this curse of weeds. They may have also known this affected their blessing from God and allowed sin into the world and was a pivotal event for humanity. Even though the Genesis 1 humans had not sinned as Adam had, Adam's single act of disobedience condemned the entire human race by spreading sin to all other humans. (Romans 5:12-17; 1 Timothy 2:14)

God blessed the male and female humans at the end of the sixth day in Genesis 1, saying everything He made was "very good," but He punishes Adam and Eve with hardships for both them and other humans. Humans aren't blessed again until after Noah's flood, when this curse of weeds is lifted.

Adam and Eve would now die, so they would reproduce. Adam then called the woman Eve because she was the "mother of all living", or "source of life." Jesus, the future descendent of Eve, was foreordained to bring about the bruise to Satan's authority over death and become the bread of life, the reason and source for those of us who will live eternally, both spiritually and physically. (John 6:35,48; Romans 16:20)

Genesis 3:20-24: And Adam called his wife's name Eve, because she was the mother of all living. Also for Adam and his wife the Lord God made tunics of skin, and clothed them. Then the Lord God said, 'Behold, the man has become like one of Us, to know good and evil. And now, lest he put out his hand and take also of the tree of life, and eat, and live forever'— therefore the Lord God sent him out of the garden of Eden to till the ground from which he was taken. So He drove out the man; and He placed cherubim at the east of the garden of Eden, and a flaming sword which turned every way, to guard the way to the tree of life.

Some argue Genesis is literal and some argue it is abstract and symbolic.[10] I have found it is both. The fruit from the tree of good and evil, the serpent and death, these were real, physical, tangible things, and also symbolic of much more. God, in His manifold, multifaceted wisdom, tells us what literally happened both on a small physical level, and also provides a foundation for much broader applications and understanding.

The serpent was real and was also Satan. (Revelation 12:9)

The fruit was real and also symbolizes the release of the knowledge of good and evil, when Satan obtained the power of death and some control over Earth, Hades and Paradise.

Physical death became imminent for Adam and Eve, but this death also symbolizes the spiritual death Satan gained the power of over human souls after they died. Later, Jesus died on the cross, abolishing this spiritual death and brought spiritual life through the gospel, and the future hope of physical immortality. (2 Timothy 1:9-10)

God created the precise physics that governs our Universe, made five amazing Eras to prepare for ours, foreordained humanities' way to eternal life, created our Era, provided humanities farm animals and crops, intervened with violent angel-human giants and the barbaric violence of the Canaanites, provided modern civilization's foundational moral code, provided a new guiding principle to love God and others as ourselves, died for us and offers eternal life as a free gift to any who accepts it, and will someday refabricate superbodies for all who accept this gift.

5

Cain, Seth and Jesus

Cain

Who was the Lord concerned about that might kill Cain?

Who did Cain build a city for?

Genesis 4:1-18 Now Adam knew Eve his wife, and she conceived and bore Cain, and said, "I have acquired a man from the Lord."

Then she bore again, this time his brother Abel. Now Abel was a keeper of sheep, but Cain was a tiller of the ground. And in the process of time it came to pass that Cain brought an offering of the fruit of the ground to the Lord. Abel also brought of the firstborn of his flock and of their fat. And the Lord respected Abel and his offering, but He did not respect Cain and his offering. And Cain was very angry, and his countenance fell.

So the Lord said to Cain, "Why are you angry? And why has your countenance fallen? If you do well, will you not be accepted? And if you do not do well, sin lies at the door. And its desire is for you, but you should rule over it."

Now Cain talked with Abel his brother; and it came to pass, when they were in the field, that Cain rose up against Abel his brother and killed him.

Then the Lord said to Cain, "Where is Abel your brother?"

He said, "I do not know. Am I my brother's keeper?"

And He said, "What have you done? The voice of your brother's blood cries out to Me from the ground. So now you are cursed from the earth, which has opened its mouth to receive your brother's blood from your hand. When you till the ground, it shall no longer yield its strength to you. A fugitive and a vagabond you shall be on the earth."

And Cain said to the Lord, "My punishment is greater than I can bear! Surely You have driven me out this day from the face of the ground; I shall be hidden from Your face; I shall be a fugitive and a vagabond on the earth, and it will happen that anyone who finds me will kill me."

And the Lord said to him, "Therefore, whoever kills Cain, vengeance shall be taken on him sevenfold." And the Lord set a mark on Cain, lest anyone finding him should kill him.

Then Cain went out from the presence of the Lord and dwelt in the land of Nod on the east of Eden. And Cain knew his wife, and she conceived and bore Enoch. And he built a city, and called the name of the city after the name of his son— Enoch. To Enoch was born Irad; and Irad begot Mehujael, and Mehujael begot Methushael, and Methushael begot Lamech.

If Cain's wife was a Genesis 1 human, then it was not incest.

Cain's punishment as a fugitive affects his relationship with others, and his immediate response is fearing people will kill him. God validates Cain's concern by having mercy and putting a mark on Cain, so they won't kill him for his crime of murder. Who might find and punish Cain for his crime of murder? Seth wasn't born yet.

There are no indication Adam and Eve might kill their only son. This is not a "my parents are going to kill me for this" expression. Cain's immediate concern is some people will kill him after they find out what he did. Whenever one commits a crime, their immediate reaction is often their most honest one, before they have a chance to think things over and filter what they say. It is unlikely that Cain immediately became very concerned about future generations finding him many years later and killing him.

If it wasn't Adam and Eve, and if Cain didn't immediately speculate about what future generations might do to him many years later, then who did the Lord agree might kill Cain? Genesis 5:4 states that Adam had additional sons and daughters who were born *after* Seth was born.

What if it was argued that Genesis was written out of order, and that nameless brothers were born *before* Seth was born?

It is emphasized throughout scripture that the first-born males carry on the royal lineage leading to Jesus. Exceptions to this are almost always documented, such as Seth, since Cain disqualified himself by murdering Able. Other famous Bible stories revolving around the extenuating circumstances disqualifying the biological first-born are— Isaac v. Ishmael; Jacob v. Esau; Judah v. Reuben et al.; Perez v. Zerah; David v. Eliab et al; and Solomon v. Amnon, Absalom and Adonijah et al. If another male was born prior to Seth, wouldn't scripture explain why he was disqualified as the first-born male to carry on the genetic lineage that led to Jesus, just as it does in these other well-documented accounts throughout the Old Testament? Judah's is summarized in 1 Chronicles 5:1-2. The importance of the first-born is discussed in Exodus 13:2, Deuteronomy 21:17, Colossians 1:15 and Revelation 1:5. Jesus needed to descend from Adam's qualified first-born male to be of royal descent, to abide by the laws and written requirements of the spiritual realm. (Colossians 2:14-15; Revelation 5)

Seth is born next and carries this first-born lineage for Jesus. Arguing there was another male born *before* Seth is an argument against the first-born lineage of Jesus Christ. Even *if* another male was born before Seth, his children would have to quickly and incestuously produce enough offspring to pose a threat for killing Cain in less than 100 years. Therefore, it is scripturally consistent that Cain was worried he would be killed by Genesis 1 humans. The Lord validates Cain's concern.

Cain's concern they would kill him for murder gives insight into the social behaviors of Genesis 1 humans. Genesis 1 humans had God's blessing on them and may have had enhanced amiable social skills and self-control with no tolerance for murder, which they would punish with death. But these humans respected God's protective mark on Cain because they knew the Lord would punish them for not obeying. Cain then built Earth's first city of Enoch for them. Does the city of Enoch frame Gilgamesh's celebrated first city of Uruk, and hence Babylon, as being founded by a murderer, and that even in civilization's earliest origins cities had co-existed with violence from the very beginning? Does Enoch reflect on the ancient cities of Eridu or Ur? The Babylonians viewed Eridu as their first city, where their god Enki gave humanity the first elements of civilization, and from which their goddess Inanna took these elements to Uruk. Abraham was from Ur.

Genesis 4:19-24 Then Lamech took for himself two wives: the name of one was Adah, and the name of the second was Zillah. And Adah bore Jabal. He was the father of those who dwell in tents and have livestock. His brother's name was Jubal. He was the father of all those who play the harp and flute. And as for Zillah, she also bore Tubal-Cain, an instructor of every craftsman in bronze and iron. And the sister of Tubal-Cain was Naamah.

Then Lamech said to his wives, "Adah and Zillah, hear my voice; Wives of Lamech, listen to my speech! For I have killed a man for wounding me, even a young man for hurting me. If Cain shall be avenged sevenfold, then Lamech seventy-sevenfold."

God said anyone killing Cain would be punished seven times, but now Lamech argues he should be avenged seventy-seven times more than Cain, as if he were kind of special for killing a man for wounding him. This lack of punishment for murder likely contributed to the extreme violence later on in the time of Noah. God instituted the death sentence for murder *after* Noah's flood to help curb violence.

Cain is noted to be the father of those who dwell in tents and have livestock, the father of all those who play the harp and flute, and of all those who are an instructor of every craftsman in bronze and iron, who are now farm managers, inventors, builders and computer designers. This is a primitive way of saying humans are endowed with the capabilities to make smart phones and be rock stars. Since we have people skilled in these areas, some of Cain's genetics, and the genetics of Genesis 1 humans, passed on through Noah. Adam's descendants had to have married the descendants of Cain. Then, Noah, a genetic descendent of Eve, survives the great flood, making Eve the mother of all living in Noah's area, and allows Cain's genetics and Genesis 1 humans to be passed on. There were also Genesis 1 humans living in other reaches of the globe that these Mesopotamian inhabitants had no knowledge of.

Seth

Genesis 4:25-26 And Adam knew his wife again, and she bore a son and named him Seth, "For God has appointed another seed for me instead of Abel, whom Cain killed." And as for Seth, to him also a son was born; and he named him Enosh. Then men began to call on the name of the Lord.

Seth obtained the first-born rights since Cain murdered Able. If other males were born prior to Seth, wouldn't scripture explain why they were disqualified as the first-born?

Mankind's Blessing and Adam's Curse

Genesis 5:1-3 (NASB)

This is the book of the generations of Adam.

In the day when God created man, He made him in the likeness of God. He created them male and female, and He blessed them and named them Man in the day when they were created.

When Adam had lived one hundred and thirty years, he became the father of a son in his own likeness, according to his image, and named him Seth.

God makes it clear the Bible and its events are about Adam's genealogy and not about other humans living at the time. Adam was made directly by God, and so was a first-generation son of God,.

God created and blessed humans during the sixth age in Genesis 1.

There is a Hebrew play on words in Genesis with "adam"— The Hebrew "adam" can be translated as the species of mankind, a man, men, humans, or "Adam" can refer to the individual Adam, or to the city of Adam. Some older translations substitute "Adam" instead of "Mankind" in Genesis 5:2, which adds to the confusion.

Were Adam and Eve ever blessed by God? God punishes them for disobeying by driving them out of the garden to die and then guards the tree of life. They must work hard and the ground is cursed. After Cain murders Able, humanity become increasingly violent until they are destroyed by God in a flood. God then lifts His curse on the ground and restores His blessing on Noah's descendants.

Death and Cain's Genetics

Did Adam's age in Genesis begin from the year of his formation or from when he was prevented from eating of the tree of life? God saying "you will surely die" meant that Adam would now age and physically die, but also alludes to a spiritual power of death and separation from God, because it was then that Satan gained control over human souls. Adam's soul will now go to either Paradise or Hades. There is more on this in the God's Master Plan chapter, "The Soul and Super Body."

Adam and Eve already had emotions, since Eve was tempted by discontent and desired the fruit to become wise. God said Adam will surely die, but there is no reference to other living creatures dying. Adam knew what death was, so physical death already existed. Eve was told she would have much more pain in childbirth, so she knew what pain was. Were small creatures made with camouflage to evade predators? Were there five Eras before ours with death and extinction?

Although animals have emotions as humans do, they are never mentioned as having spirits or being resurrected.

I am the way, the truth, and the life – Jesus

The First and Last Man

Paul said just as Earth is different from other planets and the sun differs from the moon, and stars differ from each other, so it is that after the resurrection of the dead the earthly bodies for believers will be physically raised in strength and made into superior new superbodies.

Paul writes, "There is a natural body, and there is a spiritual body. As it is written, "The first man Adam became a living being." The last Adam became a life-giving spirit." Believers have a physical body like Adam, and become spiritually reborn in Christ so that someday we will have a new physical body like Jesus, the heavenly Man. (1 Corinthians 15:40-49)

Adam is the first man who brought sin and death into the world.

Jesus is the last man who took care of the sin requirement that we are all under because of Adam. Jesus is the second man who loves all humans and died to bring them life and immortality. Jesus makes this vital physical resurrection of the dead possible. Someday soon, our new bodies will be like His new resurrected body.

When Paul refers to Adam as the first man, is he referring to Adam's genealogy to Jesus, or to the genealogy of the human race?

Adding who existed helps clarify this is genealogy:

"The first man Adam who existed became a living being"

"The last Adam who existed became a life-giving spirit"

Jesus is not the last man who existed. It's about Adam's genealogy.

God formed Adam from dust, and Eve from Adam, making them genetically separate from all other humans and the first in their lineage. We first get an inkling of this significance in Genesis 5:1, when it stated, "This is the book of the genealogy of Adam." The Bible then goes on to detail the lives of Adam's descendants leading to Jesus.

The genealogy of Jesus is important because of the laws and written requirements in the spiritual realm that had to be met. Adam was a son of God because he was formed directly by God. The back story, we are told later, is that Satan was a son of God and obtained the "sting of sin" from Adam's disobedience to gain control over physical human death and human souls. It was required that another son of God, who was also a direct descendant of Adam, fulfilled the requirements of the spiritual realm in order to conquer Satan's realm of death on Earth. Satan's dominion of death had spread to all other humans, reigning even over those who had not sinned like Adam had. Jesus had to overcome Adam's sin, the sting of death.

Jesus, formed directly by God's Spirit, was the son of God who ended Satan's control over death. The essence of the Bible is telling of why Jesus had to die for the sins of humanity, and of God's interactions with the genealogy of Adam who led to Jesus. The spirit of the Bible is the testimony, or the clear witness, of Jesus. Jesus died so we can become spiritually reborn children of God if we accept His gift of eternal life and God's Spirit, and will someday receive new refabricated superbodies. There is more on this in the God's Master Plan chapters, "The Conflict in the Heavens," "The Book of Life" and "The Spirit and Super Body." (John 1:1-3, 3:1-21, 11:25; Romans 5:12-17; 1 Corinthians 15:21-23, 41-43, 54-57; Colossians 2:13-14; 1 John 3:2; Revelation 5; 19:10)

It's easier to play it safe and go with the status quo of what our friends think, but human souls are the real issue. Science now has evidence to support humanity was way before Adam, and so they see Adam, and Jesus, as just a myth. Laziness is destructive.

Counterpoints

If Adam and Eve were made after the humans of Genesis 1 were created, then the rest of the Bible must be consistent. Paul's statement that Adam was the first man, and Adam's statement that Eve was the mother of all living are, I think, the two strongest counterpoints. Adam called Eve the source of all living in response to God telling Adam that Eve would birth his descendants. But God is also, I think, referring to Jesus, the future descendent of Eve, becoming the source for eternal life.

So, Genesis is found to be consistent with anthropological historical and geological findings, but these two new inconsistencies arise.

Paul told the philosophers at the Areopagus that God made all humans in all nations, and we are all of one blood. Although planetwide human genetic testings have confirmed this, God *can* make genetically related humans out of clay or stones. Paul doesn't say we are all Adam's offspring, but rather we are all God's offspring. *If* Adam was the very first created human, then he would be a father of the human race and we would all be his offspring, just as Abraham is a father of many nations. But this title of a father, or that we are all Adam's offspring, is never attributed to Adam, but just that he is the first man in the genealogy leading to Jesus. Malachi 2:10 states, "Have we not all one Father? Has not one God created us?" God made all humans, and Jesus died for all of humanity, not just for Adam's descendents.[7,70,71]

(Matthew 3:9; John 3:16; Acts 17:26)

Jesus said marriage began with the Genesis 1 humans, "From the beginning of the creation, God 'made them male and female.' 'For this reason a man shall leave his father and mother and be joined to his wife, and the two shall become one flesh'." The Greek "ktisis" is translated here as "creation" and means "the original formation." Jesus directly references the creation of humans on the sixth day, not Eve's formation from Adam's rib. (Mark 10:6-8)

Can angels who are unrelated to Adam sin? Yes. (2 Peter 2:4)

Do humans who are unrelated to Adam's sin? Yes. (Romans 5:12-14)

Do humans who are unrelated to Adam need God? Yes. (Acts 17:30)

The issue is our relationship with the Lord, not who we are related to. Since translators historically believed Adam and Eve were created on the sixth day, and translated accordingly, some passages need a bit of unraveling to determine what the original text actually indicates.

Sin, Canals, and Building

Adam's action spread sin to all of humanity, but God made something very good come from his genealogy with Jesus. This compares to God's choice of Bathsheba to be Solomon's mother, of bringing some good from King David's sinful actions.

Adam's descendants introduced sophisticated canal building, metallurgy, and building abilities for modern civilizations, and murder, but Adam also allowed Satan to gain control over Hades and Paradise with the power of death through Adam's single act of disobedience. This judgment of sin spread internationally to all humans in Satan's planetary reign of death. The son of God Adam disobeyed God, so his descendant, the son of God Jesus, paid the price for Adam's sin with His own death. They were sons of God because they were made by God. (Romans 5:12-17)

Two riveting books by Nik Ripken detail how God's Spirit is working with humans internationally; *The Insanity of God: A True Story of Faith Resurrected*, and his follow up book, *The Insanity of Obedience: Walking with Jesus in Tough Places.* His first book details his journey and provides the circumstances and details behind the more summarized, and fascinating, findings in his second book.

As a blind man has no idea of colors, so have we no idea of the manner by which the all-wise God perceives and understands all things
- Isaac Newton

Love v. Consequences

The Christian community has been slow in recognizing how not-so-recent solid and proven scientific findings impact some biblical passages. We tend to ignorantly treat sound science as threatening, but are quick to say God created it all, and the only proof we have of that is our faith.

The scientific community tends to weakly encourage moral values, portraying religion as a weak part of humanities' dark past, but some Atheists endeavor to be moralistic. Their moral code, per *www.atheists.org*, is to achieve enlightened self-interest to maximize personal gratification. This entails a desire to help others by predicting the consequences of their actions. Is this similar to Christians who fear God? Is endeavoring to care for others, considering the consequences of our actions, and enjoying the fruits of our relationships with others much different than this?

Long ago, God provided Israel with the moral code that our modern civilizations are now based on; justice systems that value mercy for the less fortunate and monetary punishments, human life and peace, that all should learn and gain wisdom, helping others in need, outlawing rape and incest, women's rights, animal rights, limited government powers, and also instructions on agriculture, treating germs and general hygiene. A good book on this is, *The Book that Made Your World* by Mangalwadi.

The Bible is criticized for the Old Testament biblical laws being excessive compared to today's standards, such as death for some sexual sins, but these laws were consistent with the Mesopotamian laws of that time, and also these biblical laws were relaxed over time. Jesus did not uphold the Mosaic law of punishment by death when the woman was caught in the very act of adultery but told her to go and sin no more. Punishment for intentional abortion is never mentioned in the Bible, but the Assyrians made it punishable by death of impaling, whether the woman was still alive or not.

After the Amorites became increasingly wicked for 400 years, God told the Israelites to destroy them, but also told the Israelites that they weren't any better. As the Israelites succumbed to these same evils, God warned them many times they needed to change before the Israelites finally began to lose control of Jerusalem around 606 BC.
(Genesis 15:16; Deuteronomy 9:4; Jeremiah 32:32-36; 2 Kings 24-25; Amos 2:9-10)

Jonah's warning caused the Assyrians to change in Ninevah, and 200 years later Daniel became an advisor in Babylon, from which came some elements of the Roman Empire and our current civilizations. English and American Law are based in part on Leviticus 18-27.

What if God hadn't influenced these civilizations? What would our civilizations be like if the Amorites and Canaanites had influenced us instead with their savage violence, prostituting of daughters, child fire sacrifices and bestialities?

Jesus went way above mere consideration of our consequences and enjoying relationships when He said to love God with all our heart, soul, strength and mind, to love our neighbor as ourselves, and to have selfless, genuine love for our spouse and children. (Luke 10:27)

We know there is a bit of digging involved here, but we just ask that you stay positive.

6

The Mesozoic Era and Noah's Flood

Planetary evidence indicates a meteorite struck Earth at the end of the Mesozoic Era and caused a vaporized water and dust cloud so massive it cloaked the entire Earth in darkness.[22,47] Genesis 1:2 seems to refer to this time when Earth was a worthless undistinguishable ruin of desolation.

God caused a local earthquake when His voice shook a mountain, but He also shook Earth and its atmosphere ***before*** humans in an earthquake so massive that it shook the entire planet's land, seas *and* atmosphere. God may have referred to the end of the Mesozoic Era when He says He will **once again** shake Earth in a future earthquake so tremendous that Earth is moved in orbit and the islands and mountains are leveled. Revelation 16:18 describes this future earthquake as "a great earthquake, such a mighty and great earthquake as had not occurred **since men were on the earth**." (It occurred before humans existed.)
(Hebrews 12:25-27; Isaiah 13:13, Haggai 2:6)

Did the Mesozoic Era have intelligent spirit-enabled beings? Humans are not God's only sentient spirit-enabled beings. There was an intelligent talking reptile in Genesis 3.
(Genesis 3:1-5; Isaiah 6:2-3; Ezekiel 1; 10; Hebrews 1:13-14; Revelation 4)

These passages may refer to the ending of the Mesozoic Era—

I saw Earth as a desolation and undistinguishable ruin
Mountains shook and hills moved swiftly under dark skies
There were no men, and the birds fled upwards to the sky
Fruitful land and cities destroyed by the Lord's fierce anger.
- Jeremiah 4:23-26

Was there a fantastic Middle-Earth Mesozoic civilization whose spirit-enabled bird beings fled to space as Earth shook?
(Genesis 1:2; Haggai 2:6; Jeremiah 4:27-30/Revelation 16:18, 18:30)

"God presides over heaven's court; he pronounces judgment on the heavenly beings: 'How long will you hand down unjust decisions by favoring the wicked? Give justice to the poor and the orphan; uphold the rights of the oppressed and the destitute. Rescue the poor and helpless; deliver them from the grasp of evil people.

But these oppressors know nothing; they are so ignorant! They wander about in darkness, while the whole world is shaken to the core.

I say, 'You are gods; you are all children of the Most High.

But you will die like mere mortals and fall like every other ruler.'"

- Psalms 82:1-7 NLV

Jesus says in John 10:34-36 that You are gods was directed at Israel leaders and judges, but does this also allude to Mesozoic Era events when God judged oppressive heavenly beings who then wandered in darkness as the whole earth shook to its core?

"Have you commanded the morning since your days began, and caused the dawn to know its place, that it might take hold of the ends of the earth, and the wicked be shaken out of it? It takes on form like clay under a seal, and stands out like a garment. From the wicked their light is withheld, and the upraised arm is broken." (Job 38:12-15)

Did God command morning angels, and cause Satan the dawn to know his place, when the wicked were shaken out of the ends of Earth and light withheld? What caused dinosaurs to become extinct?
(Genesis 1:2; Isaiah 45:18; Haggai 2:6; Revelation 16:18)

Earth's foundations were laid and will forever not be moved
A deep-water garment covering above mountains
A rebuke with your voice of thunder and they fled in fear
From high mountains to the valleys
To the place You established for them
A boundary you set so they will not return and cover Earth
- Psalms 104:5-9

Psalms 104 says a boundary was set for the waters to not return and cover Earth when God made dry land appear in Genesis 1:9.

Could 2 Peter 3:4-8 refer to the Mesozoic Era?— Some people say that all things will continue as they were from the beginning of creation, but they are willing ignorant that the heavens are from long ago, and that land stood out of Earth's water by God's word. But the world that then was overflowed with water. Now, the heavens and Earth are being reserved for a judgment of fire when ungodly people are destroyed.

The Giant Flood

The Greek word "ghay" means our planet that is of old by the word of God. It is not used in 2 Peter 2:5 and Hebrews 11:7 to describe the world of violent giants destroyed in Noah's flood, but rather the Greek "kosmos," meaning the world in a narrow or broad sense, including its inhabitants. Noah's flood wasn't planetary since it did not return to cover the earth. Mesopotamian stories of Ziusudra, Utnapishtim and Atrahasis, and ancient historians Berossus of Babylon, Heironymus the Egyptian, and Nicolaus of Damascus all told of a flood in ancient times.

The mass destruction in Noah's flood can be disturbing because when we judge God as unfair, it troubles us. Some argue God isn't fair, so there must not be a God. But God put that in humans to think He should be fair. We *know* "good" *and* "evil." Consider that for a moment, if you want to. It's impacting and may cook your noodle.

The Lord says, "Yet you say, 'The way of the Lord is not fair.' Listen now, is it not My way which is fair, and your ways which are not fair? Repent, and turn from all your wrongdoing, so your sins won't be your ruin. Throw away all the wrong things you have done against Me and get a new heart and spirit. Why should you die? I have no pleasure in the death of one who dies. Therefore turn and live!" (Ezekiel 18)

Is there anything more dangerous than
dissatisfied and irresponsible gods
who don't know what they want?
– Yuval Harari

The Giants

Genesis 6:1-4 Now it came to pass, when men began to multiply on the face of the earth, and daughters were born to them, that the sons of God saw the daughters of men, that they were beautiful; and they took wives for themselves of all whom they chose. And the Lord said, "My Spirit shall not strive with man forever, for he is indeed flesh; yet his days shall be one hundred and twenty years." There were giants on the earth in those days, and also afterward, when the sons of God came in to the daughters of men and they bore children to them. Those were the mighty men who were of old, men of **renown**.

God needed the violence of Noah's time to end, which was caused in part by hybrid human-angel giants. Modern civilizations would not thrive with these violent giants ruling Mesopotamia.

Some wandering stars, fallen angelic beings, left their place and habitation and came to Earth. They bred with human women and their offspring became famous giants. It says, and also afterward, meaning this happened *after* the flood also, such as with King Og and the giants who were destroyed by the Moabites and Israelites, and Goliath and his relatives who were killed by David and his men. God sent a clear message that He wanted these genetic perversions destroyed (Deuteronomy 2:9-11,28-37, 3:1,11,13, 9:1-2; Joshua 11:21-23; 2 Samuel 21:16-22; Amos 2:9; Jude 6).

An ancient Jewish belief says that these hybrid giants had special abilities to cure themselves and recover quickly from their injuries. This is consistent with Abaddon, the inhuman angel in Revelation, who emerges from the bottomless pit and miraculously recovers from a deadly wound to his head (Revelation 9:11, 13:12, 17:8). Men of **renown** indicates that the ancient Sumerian and Greek Mythologies were derived from these famous giants. Ptolemy I Soter's son, who may have requested the Septuagint LXX translation, claimed to have descended from Hercules, the Greek mythical hero and founder of the Argead dynasty.[58,59,60,61,62,63]

"Sons of God" are first generation beings made directly by God. Adam was a "son of God," as well as the angelic "sons of God" with access to God (Job 1:6, 38:7; Psalms 89:5-7; Luke 3:38). Believers are adopted as spiritual beings by receiving God's Spirit, and we then become spiritually re-born children of God. (John 1:12; Galatians 4:6; 1 John 3:1)

Modern Civilizations

God instructed Noah's descendants to replenish Earth. They stuck together and built a city with a high tower as a monument to make a name for themselves, but God wanted them to spread out to begin modern civilizations all over the planet with other humans—

The Lord said, "Look at what these people are beginning to do, they are united, and with one language, and now nothing they propose to do will be too difficult for them. Come, let Us go down and confuse their language so they may not understand one another's speech."

The city was called Babel because the Lord confused human language there and then scattered them. He later united those within these nations who call on His name. (Genesis 11:1-9; Zephaniah 3:8-9, Acts 2:5-21)

About 2,600 years ago the Lord told Daniel that 'people of the future will travel a lot and knowledge will increase'. This perfectly depicts our planet's current sophisticated cities with encircling jets, streaming highways, sailing yachts, and skyscrapers towering over the city sprawl receding into the distance. The genetics and ability to teach that God injected into humanity with Adam and Eve has propelled human advancement into remarkable achievements since, considering our technology, agriculture, trucks and chain restaurants, homes for Earth's billions of human inhabitants, and cargo ships carrying merchandise between continents. (Daniel 12:1; Revelation 18:17)

The Genesis 1 humans were likely amiable with enhanced social skills and self-control, but their simple lifestyles and increasingly dense populations forced a need for new abilities. Humanities problems didn't stem from these new God-given abilities given via Adam, but from Adam's sin, and Cain's murder, that spread to all of humanity.

God's solution to address the dense population issue facing humanity was to make Adam with the ability to build, govern, invent and teach, and then integrate Adam's descendants into the human gene pool—Paleoclimatological findings show the global climate began to dry and stabilize ten thousand years ago,[37,66] such as the desertification of the Sahara, caused by the tilt and wobble of Earth's Milankovitch cycle.[83] This climate change encouraged dense populations, and a need for dense agricultural methods.[38,42] The governing systems, inventions and explosion of technology emerging from Mesopotamia supports modern civilization began soon after the making of Adam and Eve.[42]

There is a time difference, since science generally says modern civilization began about 6 to 10,000 years ago,[79,80,81,82] and the Genesis genealogy account indicates it was 6 to 7,500 years ago.[12,79,80,81,82,83]

The Genesis account confirms the archeological findings of Ancient Egypt, Mesoamerican, Africa, Indus Valley Civilization, Ancient China, the Sumerians, ASPRO chronology, Paleolithic hunter-gatherers, and inventions such as the wheel, all of which place modern civilization's origins in this area.[42,79,80,81,82,83,84]

Using Genesis genealogies to accurately date Adam is unreliable— The Latin Vulgate, translated from Hebrew text around AD 405, conflicts with the Greek Septuagint LXX, translated from Hebrew text around 200 BC— it includes the Cainan in Luke 3:36 which the Latin Vulgate left out. Genealogies in Matthew and Ezra 7 selectively skip some as well, those mentioned were significant— these telescoping genealogies were written with oral traditions and are meant to have numerical symbolism. Also, Adam may have been alive for a few thousand years prior to being expelled from the garden of Eden. What is clear is that modern civilizations emerged soon after Adam.[12,42,61,85]

God Can Do It

God created the Universe and changed past Eras and can flood Earth. But, after separating land from water in the third age, did He reverse that for Noah's flood? Sound geological evidence clarifies Noah's flood was not global.[77] If God destroyed the wicked people of Noah's area, and if other parts of the world also had apocalyptic flooding near this time from rapid ice melt and atmospheric rivers, would it discredit Christ? I don't think so.

Atheists and Humanists think the claims of Noah's planet-wide flood are geologically unsound.[31] What if they are right? Could clinging to a global flood belief be a **stumbling block** for others and discredit Christ? What *is* on the line for those who earnestly contend Noah's flood was global? A selfish fear that if this is wrong, then what *can* they believe? Deeply felt emotional triggers often arise when childhood memories are threatened. If one believes Noah's flood was planetary, then arguing it was local can cast doubt on the entire Bible for them. Like people targeted in the movie *Inception*, our minds can be unwittingly snared by the imprinting that is buried deep in our subconscious mind.

Noah's Flood Investigation Summary

Endowed with agricultural expertise, Adam and Eve's descendants helped colonize Noah's area, a mild, low elevation rainless Mediterranean climate with spring-fed alpine ice run off that was perfect for growing olive trees and grapes. Noah is surrounded by violent and evil giants, genetic deviants of angels and human women. These famous giants and everyone else in their area become so evil and violent that God wants them all destroyed, except for Noah and his family. The angel-human hybrids of Sumerian and Greek mythology may have its roots from this time.

God needs to save the domesticated farm animals and local creatures in Noah's area for future generations, and so we now have billions of cows, goats and sheep. God knew these agricultural animals are direly needed for a modern, densely populated planet, and specifies Noah to take fourteen each of these important species. (Goats are not as popular in America, but most of the world's population drinks goat's milk.[72])

The ark was about 500 x 80 x 50 feet, resembling the proportions of modern cargo ships that are built to maximize storage. The ark had about 130,000 square feet. Fourteen cows, sheep, goats, two camels, and two pigs, and the many thousands of native animals and birds in ancient Mesopotamia, requires about 80,000 square feet. There was plenty of room in the ark for local animals.

Noah's account conveys an intense, life-impacting event. After building the ark for many years, and after feeding and cleaning up after a lot of animals for many months in cramped quarters with his family, as everyone he ever knew beyond his immediate family was destroyed, Noah grows some grapes, makes some wine, and gets drunk.

God saves the domestic animals and grains, destroys the wicked giants and people to end their extreme violence, and resumes His Genesis 1 blessing again on humans. God resets Earth to be a habitat for modern human civilizations to develop.

This impacting event affects Noah's earth and matches geological evidence.[77] Were tribes of Genesis 1 humans in other regions of the world also affected by floods? Possibly, but debating this minutiae detracts from the main point of Noah's account.

Eight Clues of Noah's Flood

The Human Perspective

How Noah's flood is viewed hinges on if the Hebrew word "erets" should be translated as– the earth at large, a land, country, or the ground.

Should this account be understood from Noah's perspective, who thought "erets" meant the land all around him, or from the 21st century perspective of humans with internet access who think "erets" means planet Earth was entirely flooded? Noah's account acquires a much different perspective if "land" is used instead of "earth" in this narrative. If we understand that "erets" means the land all around Noah, then we can see this was mostly from Noah's perspective.[59,60]

Noah is an ancient eyewitness describing what he saw. The "earth" is the dry land where Noah lived, and "under heaven" is the sky above him. "All flesh" is the humans and animals in the land around him. Everything Noah could see was flooded and destroyed.

This account is *very* impacting, but this impact can get *buried* in the conflict over if it was a local or worldwide flood. If it is understood that "erets" means planet Earth, then a *lot* of miracles must be invented to make it work as a planet-wide flood. So, some Christians have just discounted it as abstract poetry.[30,32] But for some, it's like saying Santa Claus isn't real, it's like a betrayal of a cherished childhood memory. A planet-wide flood just makes for a better story than a large local flood with farm animals, until archeology findings later prove otherwise.[35,41,77,78]

Noah's account reflects the Mesopotamian perspective of five thousand years ago, not the 1611 perspective of Bible translators who already knew of China and America. Were native Australians and Mesoamericans *really* buying grain from Joseph when the famine was over the face of the earth, and all countries came to buy grain? (Genesis 41:56-57)

In the long run, my observations have convinced me that some men,
reasoning preposterously, first establish some conclusion in their minds
which, either because of its being their own or because of
their having received it from some person who has their entire confidence,
impresses them so deeply that one finds it
impossible to ever get it out of their heads
- Galileo Galilei

God's Perspective

After the flood, the Lord said in His heart, "I will never again curse the ground for man's sake, although the imagination of man's heart is evil from his youth; nor will I again destroy every living thing as I have done." (Genesis 8:21)

The Lord uses the Hebrew "kol," which is translated as "every" or "everything." Then God adds, "as I have done," meaning as He did in Noah's flood.

Zephaniah 1:2-3 and 1:14-18 use the Hebrew "kol" in a future context of the entire planet being destroyed. However, Revelation 9:11-20:8 clarifies that this great day and returning of the Lord is a global judgment on the wicked, just as it was a global judgment on the wicked who were destroyed in the time of Noah.

Feeding the Animals

God told Noah to bring every living thing of all flesh *and* their food.

If the ark contained our known two million planetary living species *and* their food, and they somehow fit them all in, then Noah, his wife, three sons and their wives would need to feed 148 animals per minute for 16 hours a day to get them fed twice a week. Moving five feet per animal puts their speed at 45 thousand miles per hour, not including freshwater processing and manure clean up.

However, our current two million known living species is just a small percentage of the grand total of known species, since most of them became extinct.[22] Including extinct animals from prior Eras, and somehow fitting them in, is about 5 billion, so 188 thousand animals would need to be fed per minute, putting the speed of these eight people at 56 million miles per hour, or only about twelve percent of light speed. If the upper estimates of 50 billion total species[22] are more accurate, then this puts them moving a bit faster than light speed, at about Warp 1.

Without inventing more miracles than are in the text, how did animals cross the oceans and vast land distances from America, Australia, the Galápagos Islands and Africa, and get back to their own continent or island? How were tropical and arctic animals cared for? How could they desalinate water for millions of animals?

There was enough room in the ark for thousands of animals *and* their food in Noah's region to be cared for, including the vital farm animals.

The Evaporation

Genesis 8:4 says, "Then the ark rested in the seventh month, the seventeenth day of the month, on the mountains of Ararat." Claiming the flood covered 16,854' Mount Ararat supports a three-mile-deep version of a planetwide flood covering most mountain tops, but where did all this water go afterwards? God sent a wind to dry up the water. Evaporating three-mile-deep planetary water would increase the atmospheric pressure to about 1,305 pounds per square inch.[73] Cockroaches might survive this. If the flood was planetary, wouldn't this wind have blown Noah's ark far away before it ran aground?

If Earth *had* been nearly perfectly smooth, the entire planet's land could be covered by up to 1.6 miles of water,[74] but still leaves the dilemma of a water-covered planet being dried by wind. Also, there is no sound biblical basis for any massive upheaval that caused Earth's mountains to rise during this flood; the last planet-wide earthquake occurred *before* men were on the earth. (Revelation 16:18)

A literal translation of Genesis 6:17 is, "I am bringing floodwaters on the land to destroy from under the sky all flesh in which is the breath of life; everything on the land shall die."

Is God speaking to Noah's perspective in making sure Noah knows to bring his entire family and all of the animals in his area, or to a 21st century perspective that Noah should also gather all life in North and South America, Europe, Australia, Asia, Africa and Antarctica?

The Dove

In Genesis 8:9, the dove returns because "the waters were on the face of the whole earth." From Noah's perspective, the whole earth was as far as a dove could fly. Did this dove encircle the entire planet? It was likely an ancient domesticated rock dove, a Columba Livia that flies no more than 100 miles a day, not a modern hybrid racing pigeon. This also indicates Noah may have never traveled more than a few hundred miles from his home. Noah was waiting for the dove's return, and put out his hand for her to land on, indicating it was tame and the dove's homing instinct was known to Noah.

Some translations state "ground" instead of the whole earth, further highlighting the challenges translators face in deciding how to translate the Hebrew word "erets."

The Olive Leaf

The dove found a living olive leaf within 50 miles of the ark. Olive trees thrive in well-drained soil at low elevations and die under 3 miles of diluted sea water for a year. The NOAA site states, "At a depth of 5,000 meters the pressure will be approximately 500 atmospheres or 500 times greater than the pressure at sea level."[75] Placing 7,000 pounds per inch on soil for a year would cause planet-wide severe soil sterilization from compaction and salination for a long time. But Noah planted a vineyard just after the flood. Mesopotamia is a flat plain enclosed by these mountains with many documented major floods, the largest being about five thousand years ago. Olive trees thrive in these low evelvations.[76]

Geological Evidence and the Jemdet Nasr

If Earth's mountains all rose with a biblically unmentioned upheaval, could sea life fossils that are found far from the ocean be from Noah's flood? The evidence shows this to be a geological oxymoron—

The ark settled on the Ararat mountain range, which cuts across and rests on sedimentary rock that is from the five Eras before ours.[35,76,77,78]

The waters rose 20 feet over the hill country (the Hebrew "har" can be translated as "hill country" or "mountains"). 20-foot-deep indicates these were hills Noah floated over, and he was clearing the ground by 20 feet since his 45-foot-tall ark was floating 25 feet above the water line, consistent with modern large boats which float 40% below the water line.

There was massive flooding in areas of Mesopotamia around 3200 BC— Uruk's long-distance trade abruptly terminated, their temples and city centers were destroyed, and the population collapsed. People in those areas just left their belongings and disappeared. This was followed by the 5.2ka event with many years of drought during Earth's Piora Oscillation, from about 3200 BC to 2900 BC. In Mesopotamia this period is known as the Jemdet Nasr. As Uruk recovered, their pottery became crude because they had lost their sharp tools for carving intricate designs. It took 200 years for Uruk to again have a population of about 40,000.[86,87]

Augustine's dating puts Noah's flood at about 3200 BC. Noah lived 350 years afterwards, putting him in time with King Gilgamesh, who says he talked with Noah, aka Utnapishtim, a man of great age. It would take Noah's three sons and sixteen grandsons about 200 years to again reach a population of 40,000 in Uruk.

We need to get 'er rebuilt with a chic Mid-Ubaidian flair by 2900 BC so archeolgists will never suspect what happened here.

History

Podany states in her book *The Ancient Near East* that in 630 BC King Assurbanipal said of his abilities as a scribe, "I inspected cuneiform signs on stones from before the flood which are cryptic, impenetrable … muddled up." (p.106) King Assurbanipal established the first library in Nineveh, and is the King Asnappar who deported foreigners to resettle Samaria. (Ezra 4:10)

The Babylonian historian Berossus wrote in 300 BC that locals used Ark remnants to make amulets. The AD 400 Babylonian Talmud claims Sennacherib got a piece of Noah's ark from Mount Judi to make an idol to the "Great God who saved Noah from the flood," and then vowed to sacrifice his sons if he won in a battle, but his sons killed him first. The three ancient accounts of the Eridu Genesis, the Epic of Atra-Hasis, and the Epic of Gilgamesh, all tell of the great flood in Mesopotamia.

God Resumes His Blessing on Humans

God initially blessed Genesis 1 humans on the sixth day, then put a curse on the ground after Adam and Cain's sins, and then lifted the curse and blessed Noah and his family. God now gives the death sentence for murder to curb the arising violence after Cain's murder and Lamech's killing. Murder degrades God, because humans are made in His image—

"Whoever sheds man's blood, by man his blood shall be shed; for in the image of God He made man. And as for you, be fruitful and multiply; bring forth abundantly in the earth and multiply in it."– Genesis 9:6-7

God officially adds meat to the menu, although God's preferred diet for humans seems to be vegetarian and possibly fish. (Genesis 1 humans were given seeds, fruit was given to Adam, and at the end of Revelation fruit is given and the lions will eat grass). Along with Adam's genetics came the choice to obey, and the knowledge of good and evil which spread sin and death to all humans planet-wide, even though they had not sinned as Adam had. (Luke 24:42; Romans 5:12-17)

Ubaid 0

The Ubaidians became the Sumerians from which came Abraham. Where did Ubaidians come from? There are speculations, of course, but nobody knows. Ubaidians just sprang up and mysteriously began the first complex modern civilization in southern Mesopotamia. The earliest archeological dig at Tell El-'Oueili near Ur simply dates them from Ubaid 0 to Ubaid 5. Ubaidians begin at period "0". [88,89] Unfortunately, the Ubaidians left their mysterious reptilian humanoid "Lizardmen" carvings behind with no explanation, so we can only speculate.

Bible Fairy Tales

1) The Universe had a beginning (confirmed 1964)
2) The Universe is stretching/expanding (confirmed 1929)
3) God's folded garments are His past creations (confirmed 1800's)
4) Earth has six major Ages (confirmed 1950's)
5) Animals came before Humans (confirmed 1800's)
6) Humans subdued Earth before Adam (confirmed 1850's)
7) Modern cows, goats, sheep, and modern crops originated in Mesopotamia around Adam's time (confirmed 1900's)
8) Earth's first modern cities in Mesopotamia (confirmed 1922)
9) There was a great flood in Mesopotamia (confirmed 1928)
10) Ur is not a mythical city (confirmed 1922)
11) Pilate was a prefect of Judea (confirmed 1961)
12) Jesus was a real living person (confirmed 1800's)

Long before the Lord was born as Jesus, He walked in Eden with Adam and Eve, spoke to Noah and ate Abraham's food. He is the Angel of the Lord who wrestled with Jacob, held onto Moses, spoke to all of the Israelites, burned Gideon's offering with His staff, and ascended up as the Man of God in the flame of Manoah's offering on the altar.
(Genesis 3:8, 4:16, 18:1-16; 48:15-16; Exodus 3:1-14; Judges 2:1-4, 6:11-23, 13:3-23; Isaiah 9:6-7; John 8:58; Luke 22:27; Acts 7:30-36)
The Lord is the Word who stood near Samuel and touched Jeremiah's mouth. Jesus is the Word; the I AM from the beginning who *is* God.
(1 Samuel 3:6-10; Jeremiah 1:1-9; John 1; Luke 10:17-20)
The Lord is far above all other gods, heavenly beings, and sons of God.
(Exodus 12:12, 15:11, 18:11, 20:3; Psalms 81:1; 97:9; Job 1:6, 38:7; Revelation 5:6-14)
(Set to Music as "Here on Earth" on my *Strong and True* album.)

Psalms 90 The Song of Moses

Lord, You are where I live
Before the mountains were raised
Before You formed Earth with all of its inhabitants
In time out of mind
From eternity past to eternity future You are God
But we humans turn back to dust
Yes, return, You say
A thousand years for You is like last night
We humans are swept away as in a dream
But in the morning light we sprout up like new blades of grass
In the morning we blossom and flourish
But at eventide we wither and die
We dry out under your anger
And are overwhelmed by Your wrath
You place our wrongdoings before You
Our secrets in the light of Your face
All our days pass in Your wrath
And we spend our days as a tale that's been told
Our years are seventy, if we are strong than eighty
We labor in vain but are so soon cut off and we fly away
Who knows the power of Your anger
How fearsome is Your wrath
Our days are numbered, so teach and bring our hearts to wisdom
Return to us, oh Lord, do not delay, pity us
Satisfy us in the morning with Your kindness
So we are happy and glad all of our days
We see evil, but let us see Your good on us and on our children
Let the beauty of our Lord God be on us and on what we do
Yes
What is in our power to do
Let it be upright

"Psalms 90 The Song of Moses" is on my *Psalm Pearls* album.

References

1 *Traditional Christian Battlegrounds.* www.badnewsaboutchristianity.com/ea0_trad.htm.

2 *Papal Condemnation of Galileo.* June 22, 1633. www.famous-trials.com/galileotrial/1012-condemnation.

3 *A Flat Earth, and other Nonsense.* creation.com/refuting-flat-earth.

4 *NASA Spacecraft Discovers new Magnetic Process in Turbulent Space.* May 9, 2018. www.nasa.gov/feature/goddard/2018/nasa-spacecraft-discovers-new-magnetic-process-in-turbulent-space.

5 *Magnetic Reconnection in Earth's Magnetosphere.* sci.esa.int/cluster/51744-magnetic-reconnection-in-earth-s-magnetosphere/.

6 *Galileo and the Inquisition.* galileo.rice.edu/bio/narrative_7.html.

7 Stoeckle, Mark Y., and David S. Thaler. "Why Should Mitochondria Define Species?" *Human Evolution* (Vol. 33 - n. 1-2 (1-30) – 2018). Jones, Andrew. *Humans and Animals Are (Mostly) the Same Age?* June 8, 2018. evolutionnews.org/2018/06/humans-and-animals-are-mostly-the-same-age/. *Sequencing Y Chromosomes Resolves Discrepancy in Time to Common Ancestor of Males V. Females.* science.sciencemag.org/content/341/6145/562.

8 Grün, Rainer. *Direct Dating of Human Fossils.* onlinelibrary.wiley.com/doi/10.1002/ajpa.20516/abstract.

9 Wiens, Roger. *Radiometric Dating - A Christian Perspective.* www.asa3.org/ASA/resources/Wiens.html.
Bailey, David. How reliable are geologic dates? www.sciencemeetsreligion.org/evolution/reliability.php.

10 *Report of the Creation Study Committee.* pcahistory.org/pca/studies/creation/report.html.

11 Walker, Tasman. *A Christian Response to Radiometric Dating.* www.creation.com/images/pdfs/other/5292wiens_dating.pdf.

12 *Bible Timeline.* www.answersingenesis.org/bible-timeline/.

13 Marshall, Michael. *Timeline: The evolution of life.* www.newscientist.com/article/dn17453-timeline-the-evolution-of-life.

14 *Becoming human: A timeline of human evolution.* www.newscientist.com/article/becoming-human/.

15 Singer, Emily. *How Dinosaurs Shrank and Became Birds.* www.quantamagazine.org/how-birds-evolved-from-dinosaurs-20150602.

16 Darwin, Charles. *The Origin of Species.* Chapter 6: Difficulties on Theory. www.talkorigins.org/faqs/origin/chapter6.html.

17 Montanari, Shaena. *Four Famous Transitional Fossils That Support Evolution.* forbes.com/sites/shaenamontanari/2015/11/17/four-famous-transitional-fossils-that-support-evolution/#2564ee622d8d.

18 Bailey, David. *Do gaps in the fossil record present serious difficulties for the theory of evolution.* www.sciencemeetsreligion.org/evolution/fossils.php.

19 Simmons, Geoffrey. *Billions of Missing Links: A Rational Look at the Mysteries Evolution Can't Explain* and *What Darwin Didn't Know.*

20 Raup, David. *Extinction: Bad Genes or Bad Luck?* www.nytimes.com/2015/07/16/us/david-m-raup-who-transformed-field-of-paleontology-dies-at-82.html.

21 *Comparing the human and chimpanzee genomes: Searching for needles in a haystack.* genome.cshlp.org/content/15/12/1746.full#sec-2.

22 Madigan, Michael T., et al. *Brock Biology of Microorganisms.* 15th ed. Boston: Pearson, 2018.

23 Introduction to Quantitative Genetics. Falconer, Douglas S., and Trudy F. C. Mackay. *Introduction to Quantitative Genetics*. 4th ed. Harlow, UK: Pearson, 1996.
24 How to preserve fleeting digital information with DNA for future generations www.acs.org/content/acs/en/pressroom/newsreleases/2015/august/dna-storage.html.
25 www.dnacenter.com/blog/dna-the-hard-drive-of-the-future/ singularityhub.com/2015/02/20/worlds-data-could-fit-on-a-teaspoon-sized-dna-hard-drive-and-survive-thousands-of-years.
26 Stoner, Peter W. *SCIENCE SPEAKS*. sciencespeaks.dstoner.net/.
27 *The Geologic Time Scale*. scienceviews.com/dinosaurs/geologictime.html. *Geological Time Scale*. www.ucmp.berkeley.edu/help/timeform.php.
28 *Solomon Asch conformity experiments*. www.age-of-the-sage.org/psychology/social/asch_conformity.html. *The Talk Origins Archive*. www.talkorigins.org/indexcc/list.html#CE. *Project Steve*. ncse.ngo/project-steve.
29 Swift, Art. *In U.S., Belief in Creationist View of Humans at New Low.* May 22, 2017. news.gallup.com/poll/210956/belief-creationist-view-humans-new-low.aspx.
30 *Why Are People So Upset About What Gungor Said?* www.crosswalk.com/blogs/christian-trends/gungor.html.
31 Sommer, Joseph. *Some Reasons Why Humanists Reject the Bible.* americanhumanist.org/what-is-humanism/reasons-humanists-reject-bible/.
32 *ETHICAL CONUNDRUMS* www.theguardian.com/notesandqueries/query/0,5753,-19339,00.html.
33 Herzing, Denise and Johnson, Christine. *Dolphin Communication and Cognition: Past, Present, and Future* (pp 3-13). The MIT Press. *How Intelligent Are Whales And Dolphins.* us.whales.org/whales-dolphins/how-intelligent-are-whales-and-dolphins/.
34 Dure, Beau. "Flat-Earthers are back: 'It's almost like the beginning of a new religion'". *The Guardian.* CBD, The "Flat Earth Radio Station", Popular Communications. June 1986. p. 31-34. www.americanradiohistory.com/Archive-Popular-Communications/80s/Popular-Communications-1986-06.pdf.
35 Z. R. Beydoun. *Arabian Plate Hydrocarbon Geology and Potential – a Plate Tectonics Approach, American Association of Petroleum Geologists Studies in Geology* 33. (1991) (p 49-60).
36 Alroy, John. *A Multispecies Overkill Simulation of the End-Pleistocene Megafaunal Mass Extinction* pdfs.semanticscholar.org/169d/f12bcba2fe44d8e41c3fdc885944a7e24afa.pdf.
37 Biello, David. "What Thawed the Last Ice Age?" *Scientific American*, 4 Apr. 2012, www.scientificamerican.com/article/what-thawed-the-last-ice-age/.
38 Harmon, Katherine. *Humans feasting on grains for at least 100,000 years.* blogs.scientificamerican.com/observations/humans-feasting-on-grains-for-at-least-100000-years/.
39 Janick, J. *The Origins of Fruits, Fruit Growing, and Fruit Breeding*. 2005.
40 "DNA Traces Cattle Back to a Small Herd Domesticated around 10,500 Years Ago." *Phys.org - News and Articles on Science and Technology*, phys.org/news/2012-03-dna-cattle-small-herd-domesticated.html.

41 Springer. "Genetic Diversity, Evolution and Domestication of Wheat and Barley in the Fertile Crescent". researchgate.net/publication /257614000_Genetic_Diversity_Evolution_and_Domestication_of_Wheat_an d_Barley_in_the_Fertile_Crescent.
42 *The Development of Agriculture.* www.nationalgeographic.org/news/development-agriculture/
43 *The Origin and Genetic Variation of Domestic Chickens.* www.ncbi.nlm.nih.gov/pmc/articles/PMC2873279/.
44 *Does the Hebrew Word "Yom" (Translated "Day") in Genesis 1 Mean a 24-Hour Day or Something Else?* factsandfaith.com/does-the-hebrew-word-yom-translated-day-in-genesis-1-mean-a-24-hour-day-or-something-else/.
45 Kahneman, Daniel. *Thinking, Fast and Slow.*
46 Hawking, Stephen. *The Beginning of Time.* www.hawking.org.uk/the-beginning-of-time.html.
47 Alvarez, L.W.; Alvarez, W.; Asaro, F.; Michel, H. V. (1980). "Extraterrestrial cause for the Cretaceous–Tertiary extinction". *Science.* 208 (4448): 1095–1108.
48 *Were the three days and three nights that Jesus was in the grave a full 72 hours?* bible.org/question/were-three-days-and-three-nights-jesus-was-grave-full-72-hours.
49 *What Is a Symbiotic Relationship?* sciencing.com/symbiotic-relationship-8794702.html.
50 *Aves:Fossil Record.* ucmp.berkeley.edu/diapsids/birds/birdfr.html.
51 Fields, Helen. "Dinosaur Shocker." *Smithsonian.com*, Smithsonian Institution, 1 May 2006, www.smithsonianmag.com/science-nature/dinosaur-shocker-115306469/.
52 *The Expanding Universe.* skyserver.sdss.org/dr1/en/astro/universe/universe.asp.
53 *God, Design, and Fine-Tuning*, Robert Collins citeseerx.ist.psu.edu/viewdoc/download?doi=10.1.1.470.7166&rep=rep1&type=pdf
www.newscientist.com/article/mg20227123-000-gravity-mysteries-why-is-gravity-fine-tuned/
54 Rich Deem. *Quotes from Scientists Regarding Design of the Universe.* godandscience.org/apologetics/quotes.html.
Schaefer, Henry. *Scientists and Their Gods.* leaderu.com/offices/schaefer/docs/scientists.html
55 *The Sacred Ibis debate: The first test of evolution.* journals.plos.org/plosbiology/article?id=10.1371/journal.pbio.2005558.
Georges Cuvier (1769-1832). ucmp.berkeley.edu/history/cuvier.html
Essay on The Theory of the Earth. Baron G. Cuvier. http://aughty.org/pdf/on_theory_earth.pdf.
56 Schorsch, Ismar. "Meaning in the Torah's Layout." www.jtsa.edu/torah/meaning-in-the-torahs-layout/.
57 Just, Felix. "Old Testament Statistics Compared". www.catholic-resources.org/Bible/OT-Statistics-Compared.htm.
58 Sefaria Library. "Frequently Asked Questions." Sola Group. www.solagroup.org/articles/faqs/faq_0032.html.

59 "Genesis 1:1-3 and the Literary Boundary." Scribd. www.scribd.com/document/740537201/Genesis-1-1-3-and-the-Literary-Boundary?utm_.
60 *Codex Alexandrinus* (Genesis 1:1). Masoretica. www.masoretica.org/?book=Genesis&chapter=1&manuscript=alexandrinus&verse=1&utm_.
61 Smith, Henry B Jr. Case for the Septuagint's Chronology in Genesis 5 and 11. biblearchaeology.org/research-articles/the-case-for-the-septuagints-chronology-in-genesis-5-and-11/.
62 *Ancient Jewish History: Hellenism.* www.jewishvirtuallibrary.org/hellenism-2.
63 Eastman, David. *Alexandria.* bibleodyssey.org/places/main-articles/Alexandria.
64 *Changes to the KJV since 1611: An Illustration.* bible.org/article/changes-kjv-1611-illustration.
65 Barnola, et al. "Vostok Ice Core Provides 160,000-Year Record of Atmospheric CO2". *SAO/NASA ADS: ADS Home Page*, 1 Oct. 1987, adsabs.harvard.edu/abs/1987Natur.329.408B.
66 *Holocene warming in western continental Eurasia driven by glacial retreat.* www.readcube.com/articles/10.1038/ngeo2953.
67 *Agriculture and parting from wolves shaped dog evolution, study find.* phys.org/news/2013-02-agriculture-wolves-dog-evolution.html.
68 *Horse Evolution Over 55 Million Years.* chem.tufts.edu/science/evolution/HorseEvolution.htm.
69 *Researchers pose revolutionary theory on horse evolution.* July 25, 2018. sciencedaily.com/releases/2018/01/180125135537.htm.
70 Grobler, Herman. *Did God create all nations out of one blood, or one man? (Acts 17:26).* bibledifferences.net/2014/11/22/106-blood-or-man/.
71 *The Textus Receptus.* www.skypoint.com/members/waltzmn/TR.html.
72 Young W. Park, George F.W. Haenlein: *Handbook of Milk of Non-Bovine Mammals* (2006).
73 *Venus compared to the Earth.* www.ajax.ehu.es/VEX/Venus.Earth/Venus.Earth.html.
74 *If the Earth were a smooth speroid, how deep would the ocean be?* earthscience.stackexchange.com/questions/7446/if-the-earth-were-a-smooth-spheroid-how-deep-would-the-ocean-be.
75 *Water Pressures at Ocean Depths.* www.pmel.noaa.gov/eoi/nemo1998/education/pressure.html.
76 Neyman, Greg. *Where was the Flood of Noah?* www.oldearth.org/articles/flood_location.htm.
77 Six bad arguments from Answers in Genesis. geochristian.com/2009/04/25/six-bad-arguments-from-answers-in-genesis-part-1/.
78 Generation of Arc and Within-plate Chemical Signatures in Collision Zone Magmatism: Quaternary Lavas from Kurdistan Province, Iran, *Journal of Petrology*, Volume 54, Issue 5, 1 May 2013, Pages 887–911).
79 *Institute for the Study of the Origins of Civilization.* bu.edu/cgs/citl/institute-for-the-study-of-the-origins-of-civilization/.
80 *Civilization.* www.nationalgeographic.org/encyclopedia/civilization/.
81 *A Salute to the Wheel.* www.smithsonianmag.com/science-nature/a-salute-to-the-wheel-31805121/.
82 General references: revolvy.com/. and historyatlas.com/.

83 *What Really Turned the Sahara into Desert.* www.smithsonianmag.com/science-nature/what-really-turned-sahara-desert-green-oasis-wasteland-180962668/.
84 *History of Ancient Mesopotamia* www.timemaps.com/encyclopedia/ancient-mesopotamia-history/.
85 *The Origins of Civilization* www.timemaps.com/encyclopedia/origins-of-civilization/.
86 Staubwasser, Michael, and Harvey Weiss. 2006. "Holocene Climate and Cultural Evolution in Late Prehistoric–Early Historic West Asia." Quaternary Research 66 (3): 372–87. https://doi.org/10.1016/j.yqres.2006.09.001.
87 Algaze, Guillermo. 1993. The Uruk World System: The Dynamics of Expansion of Early Mesopotamian Civilization. Chicago: University of Chicago Press. (2nd ed., 2005, includes a new chapter with updated data.)
88 Huot, Jean-Louis. 1989. "Ubaidian Villages of Lower Mesopotamia: Permanence and Evolution from the 'Ubaid 0 to 'Ubaid 4 as Seen from Tell el'Oueili." In Upon this Foundation: The 'Ubaid Reconsidered, edited by Elizabeth F. Henrickson and Ingolf Thuesen, 19–40. Copenhagen: Carsten Niebuhr Institute.
Websites accessed 11/2019

Suggested Reading:

References: 1,7,10,20,28,30,31,33,35,51,55,69,77,86,88,89,92,94.
Scripture references given throughout this book are highly recommended.

Books:
Understanding the Universe by James Seaborn
Astrophysics for People in a Hurry by Neil deGrasse Tyson
Einstein by Walter Isaacson
What Darwin Didn't Know by Dr. Geoffrey S Simmons
The Book that Made Your World by Mangalwadi

Books on early Mesopotamia:
The Ancient Near East: A very Short Introduction by Podany
Civilizations of Ancient Iraq by Foster & Foster
Babylon Mesopotamia and The Birth of Civilization by Kriwaczek
Sapiens: A Brief History of Humankind by Harari
Archaeological Landscapes of the Near East by Wilkinson
Heartland of Cities by Adams
The Holocene: An Environmental History by Roberts

Online interlinear Bibles show the original Hebrew and Greek words with their assigned Strongs Concordance numbers. These Strongs numbers can then be used for reverse word searches (sometimes called Englishman Concordance) on Hebrew (Old Testament) and Greek (New Testament) text to see their context with other verses.

About the Author

Rhett Otis and his wife live in Virginia and have four children. He has a Bachelor of Science degree in organizational behavior from California State Polytechnic University, and investigates civil and white-collar fraud. He has believed in Jesus since 1973.

Books

God and Humans
God's Master Plan

Books under pseudonym Rhett Stuart

The Bishop Maker – When the Battle is for the Subconscious
How to recognize and recover from conservative Christian cults

Generation Modified
Young Adult science fiction

Music

Rhett Otis is on most streaming services
Lyrics aspire to scriptural accuracy
Spotify Playlists:
Gospel Indie Favorites
Beach Rock Gospel Indie
Locked In Gospel Indie
Easy Listening Gospel Indie
Country Gospel Indie

Bummer about the Big T and all,
I'm really gonna miss those after parties.

RHETT OTIS

Book Two

God's Master Plan

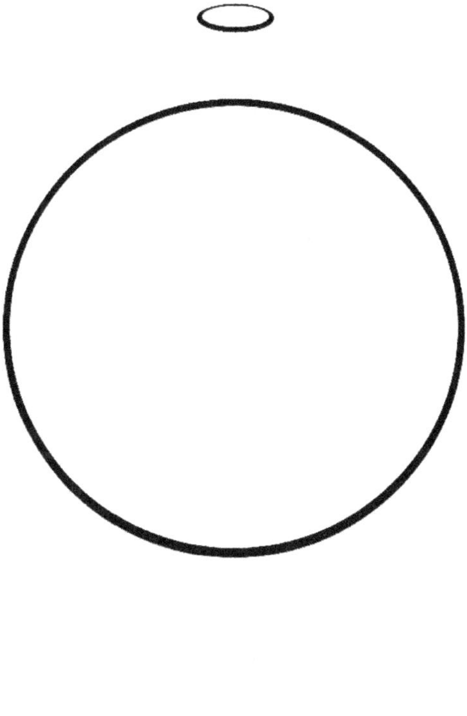

God's Master Plan

by RHETT OTIS

Summary: Findings from a holistic investigation of God's master plan for humanity and how God is dealing with the conflict with Satan in the heavens, why some names are blotted from the Book of Life, how our heart, soul and spirit interact and where they go after death, what Hades and Heaven is and why Paradise is now empty, the superbody refabrications, the Eternal Fire, what to expect in the end time events and the future mass extinctions, angelic transportation devices (ATDs), and the seventh Era yet to come. Discusses God's precepts for having an abundant life in Christ, successful relationships, marriage, and raising children.

ISBN: 978-0-9829547-9-9
Library of Congress Control Number: 2020904172
Copyright: 1-8592186131
 Missage Media - Charlottesville, Virginia

1

The Investigation

Perhaps, like me, you long to clearly understand God's masterplan for humanity, and the purpose for our sojourn here on Earth. Why does our Creator desire real loving relationships with each one of us, and why does He use even our most difficult trials to personally draw us closer to Him?

Have you wondered where human souls go, and how this changed after Jesus died on the cross? Why it's so important to keep our name in God's Book of Life? What Heaven is like, and what we will be doing there for all of eternity? How did Satan become Earth's god and the source of our problems? Why hasn't the supremely powerful God already put an end to evil? Have you suspected a much larger universal conflict is spilling chaos into our world?

My lifelong investigation into this has drawn me closer in my daily walk with God, and I hope this summary encourages you as well. God has a grand purpose for humanity, and we fit into a much greater panorama than we physically see. Seeing this big picture brings clarity to God's heart and character, and helps reveal our role in His masterplan.

An understanding of biblical concepts provides peaceful principles for strengthening personal relationships, enjoying marriage, and peacefully raising kids. Understanding how our spirit, heart and soul interact helps us understand and manage our deep internal motives, and knowing how God designed them to work together helps us to better understand ourselves, relate better with others, and make wise decisions.

Heart, Soul, and Spirit Preview

God's living and powerful Word discerns our soul's thoughts and our spirit's intents that are so closely entwined within our heart's decisions. (Hebrews 4:12)

The Hebrew "nephesh" is used for both animal and human **souls**, or **life**. The Bible doesn't say the souls of animals survive death as human souls do, but knowing they have souls helps explain why animals have emotions, such as intelligent dolphins with their enhanced self-control and social skills. Our valiant Australian Shepard feels brave and fears the unknown, is often elated to see us, and enjoys companionship. God gave animals these feelings; He gave them life, or souls, as He did to humans. There is much said about the human soul, yet scripture places more emphasis on our spirit! (Genesis 1:20,24,30; 2:7,19; 9:4)

Our "**heart**" is our will and what we use to make our decisions and believe in God. We need to diligently guard our heart because it determines what we do in our lives. (Proverbs 4:23: Mark 15:19; Romans 10:10)

Our **spirit** provides our intents and influences and urges us to do things at times. We should consider our spirit's intents and our soul's thoughts and desires, but our heart ultimately decides what we will do. Proverbs 16:32 says we need to control our spirit. God's Holy Spirit enters and dwells in us after we believe in God, and provides us with God's intentions. We decide in our heart what to do with these gentle whispers and intents from the Holy Spirit. We don't have to obey God, and if we do disobey, we can quench this influence of the Holy Spirit. The Holy Spirit connects all believers as One Spirit! In realizing this, it dawned on me why I sometimes think of someone and pray for them and then later find out they really needed prayers at that time! It's not some physic ability, it is the Holy Spirit connecting us and softly whispering to us when others need our prayers! Isn't that so cool? It's amazing. (1 Thessalonians 5:19)

The Lord also has a heart and a soul (Jeremiah 32:41), but angels and heavenly beings from God's heavenly realm are just said to have spirits (1 Kings 22:21; Ezekiel 10:17; Hebrews 1:14), so we have more in common with the Lord than with angels (Genesis 1:26-27).

The old expression "follow your heart" implies that we should pursue our soul's desires and our spirit's deeply felt aspirations. Since the human heart, biblically, is our will, we should first *guard* our heart because we use it to make our decisions, then *consider* our spirit's intents and our soul's thoughts and desires, and *then* decide in our heart what to do. As Proverbs 16:32 points out, we should control our spirit, not follow it. People who allow their spirit to control them aren't thinking things through deeply enough. When we relinquish our heart' decision-making control to our spirit it makes us vulnerable to hasty and poorly thought-out decisions. (Proverbs 4:23, 23:19-21; Malachi 2:13-16)

Kahneman points out in his book *Thinking, Fast and Slow* how we tend to instinctively think fast and consequently we intuitively come to a wrong conclusion. Daniel points out we need to think carefully and deliberately and engage our intellect so we can make better decisions. This slow, deep thinking engages our heart to consult with our spirit's intents and our soul's thoughts and feelings before we carefully make a wiser decision. For a believer, "slow thinking" includes mediation, reading God's Word and prayer. In other words, we should allow time for concentrated thought and prayer before making important decisions.

Knowing how our soul, spirit and heart work together helps us become more aware of the differences between our desires, intents, thoughts, feelings, motives and desires. The "OUR HEART", "SOUL", "SPIRIT", and "The Abundant Life" sections are almost entirely composed of woven together scripture, and for me they radiate with the Holy Spirit and God's love. I hope you will find this to be true also, and that these truths will motivate and inspire you every day—

We should be glad when our trials come, because it is through patiently enduring them that we grow in our faith and our character grows stronger. The burden Jesus places on us is light and easy, and our **soul** finds rest when we make our requests known to Him, learn to trust Him by carefully reading God's word, and walk in harmony with Him. He is the Father of our **spirit** and we are His children, and this is why He wants to help us with our painful problems; to unburden us, heal us and give us peace. The Lord comes close to those who have a humble and broken and contrite **heart**. He sends His Spirit into our hearts, and we cry, "Daddy, Father!" God's Spirit dwells in our new adopted heart to comfort, teach and bless us.

The Clouds of Heaven and White Spaceships

Was Revelation written with the ancient Greek words that best fit our *current* time perspective? Revelation 19:11-14 says when Jesus and His army return to Earth they will be sitting upon or residing within white "hippos". The Greek "hippos" meant "horse", what a conquering leader would use as transportation to enter battle. But now "hippos" refers to a barrel-shaped object. Jesus and His army will be within or upon white barrel-shaped transportation objects.

First, a great cloud conceals Jesus and His army as they descend towards Earth. The descent of this great cloud takes some time, since the rebellious rulers of Earth can see the sign of Jesus coming in these great clouds in space, and the beast angel Abaddon has enough time to gather a massive army to fight Jesus. Then, Jesus descends from this great cloud, just as He was lifted into a cloud when He ascended from Earth so long ago. (Acts 1:9-11)

When this great cloud reveals Jesus and His army, will they be returning on white horses, or be in white spaceships? Horses and chariots were the means of transportation thousands of years ago. But, if Revelation was translated with today's perspective in mind, "hippos" might possibly be translated as "Angelic Transportation Devices" (ATDs). God created physics and angels. No doubt Jesus and His angels know more about quantum physics and have fantastic transportation devices that are much more highly advanced than anything humanity has! However Jesus returns, it is going to be super cool!

Consider the science fiction movies about terrible space invaders who attack Earth. The vulnerable humans fight back and somehow defeat these space invaders. I enjoy watching science fiction movies, but find it interesting there is a common theme that these space invaders are evil. Jesus made Earth, and God owns it. Satan has just taken it over, at least in part; and we humans haven't been doing all that great of a job caring for it. Jesus will take it back and reign over Earth for a very peaceful and prosperous one-thousand years. No doubt, when Jesus returns in this great cloud, He will be regarded as an alien invader to those unbelieving humans who are still here on Earth. Maybe that is why I suspect Jesus and His army will return in white spacecrafts; it fits the profile of what humanity has been conditioned to fear and hate.

God's Orbiting City and iEarth 7 Preview

In Heaven, will we sit on clouds and play harps, or will we have upgraded superbodies and live *with* God on an amazingly renovated Planet Earth? An Earth so fantastic that *Avatar's* Pandora pales in comparison? God gave us a few clues about this—

After the seventh trumpet has sounded near the end of the tribulation in Revelation 16, the sea is ruined, the mountains are leveled and the islands have fled. If you are thinking of waiting this out on an island, consider that even if your boat withstands the massive tsunami waves from the planetary earthquake that levels all of Earth's mountains and islands, the oceans will become so polluted that everything in it dies.

God then heals our planet and Jesus reigns over planet Earth for one-thousand peaceful years. The animals become so peaceful that lions will even eat grass! With peace on Earth and long-life spans, technology may exponentially advance so rapidly that the technology of science fiction movies will seem juvenile in comparison. At the end of this utopian one-thousand years, Satan, who was locked up at the beginning of it, is released and once again wreaks havoc and turns the nations against Jesus. They surround the holy city that Jesus is in with His believers. Fire comes down from God and burns up the attacking army, and likely the entire planet; we may be like Shadrach, Meshach and Abed-Nego in the fiery furnace! (2 Peter 3:7-10; Revelation 20:9)

In Revelation 21, after our Lord's millennial one-thousand-year reign ends with this fire, God says He will now make all things new for His new Earth. Since God's last six Eras were here on Earth (as I've come to realize), and assuming that He will renovate our planet once again (which it seems to indicate), what will God's awesome seventh Era be like, His new iEarth 7? It will no longer have vast oceans, so our planet may have high water tables with vast aquafers to distribute geothermal heat and water the plants and trees as it may have in prior Eras. I picture an interlacing of gentle hills, lakes, streams, waterfalls, Jurassic Era redwoods and giant sequoias, amazing new fantastic kinds of trees, plants and animals, and new relationships with those who dwell among the stars. We may even have the ability to teleport, as Jesus could. When we look up, will we see God's new translucent gold city orbiting overhead?

When God the Father dwells with us, will His fantastic new city also serve as an orbiting interstellar gateway space city for both humans and those who dwell in the heavens, as Revelation 21:10 seems to indicate?— John is taken to a very high altitude (many translations state this is as a "great high mountain", but this Greek "oros" simply indicates that John was lifted "exceedingly high above the plains". Also keep in mind that Earth's mountains have all been leveled *before* this occurs). John observes an angel measuring God's beautiful new city as it descends towards Earth. The angel is told to measure the city, the city's gates, and the wall. The city is square- 1,400 miles by 1,400 miles, the altitude is 1,400 miles, and the wall is 200 feet. Did you catch the measurement of the gates? They are 1,400 miles high!

1,400-mile-high gates would stick out from Earth's surface like an enormous appendage! However, if this city *orbits* Earth, then it makes a lot of sense why these gates will be at an altitude of 1,400-miles. The passage never states that God's city actually comes to rest on Earth. Also, the Greek "hupos" used for this gate "height" means "altitude in the sky"! People long ago would have had difficulty understanding how a solid gold city could stay in Middle Earth Orbit. They didn't know of the Roche limits for planetary orbit. I think it is super cool that, *if* God's new city is to be 1,400 miles high in lower Middle Earth orbit with a jeweled 200-foot-thick dome in the less hazardous lower part of the Van Allen radiation belt, it does comply with our current laws of physics. It also states God will make all things new, so it doesn't necessarily *have* to comply with our current understanding of physics. The city's gates always stay open. How could this be? Movies have no difficulty depicting how space gates can stay open. God Himself will dwell there, so if He wants the gates to always stay open, or have His city in geostationary orbit, or wants to make another Universe, He will do it.

What does the Bible say about those extraterrestrials who dwell among the stars, and what interest do they have in humanity? It does say a little— Ephesians 3 explains that from the very beginning God kept His mysterious eternal plan a secret from them. Part of God's purpose of Jesus dying for our sins so those who believe can receive eternal life was to display His wisdom to these rulers and authorities of unseen places in the Universe— God reconciled all things to Himself, the things in heaven and the things on Earth. (Colossians 1:20)

There *are* extraterrestrials. It's biblical. God will reconcile those who dwell in the heavens with humanity, and it could be that God's new city will orbit Earth as a space station. Jesus said, "In my Father's house are many mansions." When Satan is finally cast out of the heavens, those who "dwell in the heavens" will rejoice— are they now living in these celestial mansions? (John 14:2; Colossians 1:20; Revelation 6:13,9:1-19,12:12)

It may be difficult to visualize quantum-based fifth-dimension extraterrestrial angels without our being biased by their humanistic portrayal in movies. I can't recall any science fiction movie where an incredible omnipotent being with genuine goodwill and love for all creates the Universe, angels, cherubim and many other things, makes six Eras of living creatures and humans on Earth, and knows them all by name. A God who allows humans to have free will and disobey Him if they choose. A God who patiently works through problems and issues over millions of years, predestines events billions of years in advance, dwells outside of time, and is all knowing.

God's realm and angels have a certain association with light—

God dwells in an unapproachable light. When the Lord spoke with Daniel, His face was as lightning and His eyes like flaming torches. When Jesus became transfigured and spoke with Moses and Elijah, His face shone like the sun and His clothes were brilliantly lit with light. The face of the angel who rolled back the stone from the tomb of Jesus was filled with a lightning-like light and had clothes as white as snow. Moses' face glowed with so much light after being in the Lord's presence that he frightened others. This indicates our future upgraded bodies will also be filled with some quantum-enabled light, a fifth-dimension energy not as we now dimly understand it in quantum physics, but an eternal spiritual lifeforce energy radiating with a kind of light that is difficult for us to physically see, or understand, from our current perspective. Jesus teleported through walls and rose into the air after His resurrection. When He appears, we will be like Him, because we will see Him as He really is, so we may also share some of these same attributes.

Facts are stubborn things;
and whatever may be our wishes, our inclinations,
or the dictates of our passion, they cannot alter
the state of facts and evidence
– John Adams

End Times Preview – Great Signs and Wonders

The Bible tells of a constitution with specific written requirements that governs all beings in the Universe, both angels and human. Long ago the humans of Planet Earth were condemned by these written requirements. God the Father, who dwells in Eternity, is called the Lord of Hosts in the Old Testament, and His Son is known as the Angel of the Lord (Zechariah 1:12). This Angel of the Lord was born as Jesus in human form to fulfil these written requirements by dying on the cross to reconcile and bring together His Heavenly family with His Earthly family.

Jesus provided a way for those humans who believe in Him and want to be on His side to later be enabled to live forever physically. Powerful, and currently unseen, angels and extraterrestrial powers can now no longer condemn humanity because Jesus disarmed them, making a universal or galactic public spectacle of them, and triumphing over them. In the future Jesus will bring universal peace, and every resident on Earth, and every resident on other planets, all of those in the Universe, will bow to Jesus. Jesus will finally reconcile His family in heaven and His family on Earth. God's golden city will descend towards Earth and, as it seems to say, orbit Earth to serve as the Lord's gateway space station for both humans and angels. Believers will then judge angels.

At the end of a great past or future war in heaven (determining if this is a past or future event is complex, considering that time itself varies, such as the time dilation of near lightspeed space travel), Michael and the angels on the Lord's side physically cast Satan and his angels out of heaven and down to Earth. Those who are dwelling in the Universe rejoice when Satan and his angels are cast to Earth, and a mighty voice says, "Woe to you of Earth, for what is about to come!"

Revelation 6 unseals an overview of how six events will impact humanity. Revelation 8 and 9 are written from a Universal perspective and announce with trumpets six end-time events with celestial origins. These events occur after our nations become perplexed by the behavior of our sun and moon, and the raging waves of Earth's oceans.
(Luke 21:25-26)

Some of the celestial events in Revelation 8 and 9 seem to involve at least two massive angelic transportation devices (ATDs):

1) An enormous storm of small meteorites flame through the atmosphere and cause massive fires on one third of Earth's land area.

2) A great mass like a very large mountain falls to Earth and destroys one third of the sea creatures and ships.

3) The massive radioactive starship? Wormwood comes burning down through the atmosphere like a torch. The debris falling onto Earth contaminates one third of Earth's fresh water. Many die from poisoning.

4) A massive object begins orbiting Earth that is so enormous it blocks out one third of the sky (the sun, moon and stars). This may describe a mammoth spaceship that is orbiting Earth.

5) The Abyss portal is opened with a key amidst dense smoke, and a vast winged army of abominable stinging creatures emerge, which could be physical incarnations of the demons that Jesus had commanded into this abyss (Luke 10:17-20). Abaddon, a highly intelligent and destructive angel, also emerges. He controls these creatures that cause tremendous pain, but not death, for five months.

6) Four angels are unbound and they lead a vast army of 200 million riders who invade Earth (possibly coming from the mammoth ATD orbiting Earth). These riders have armor with superhero-like colors of fiery red, hyacinth blue and sulfur yellow. They are in or upon barrel-shaped transportation devices with fiery fronts which emit highly toxic sulfur dioxide gas and brimstone. Located in the back of these flying devices is some malicious, cunning, and deadly technology.

Later, Revelation 17:8 explains that the "beast" who ascends out of the Abyss "was, and is not, and yet is". This must be Abaddon, since he is the destructive angel who ascends from the Abyss— The stories of Greek Mythology about their god's improprieties and affairs with human women, such as Apollon, may have originated from the angels who had offspring with human women in the time of Noah. Genesis 6 says these giant offspring were "men of renown", so well before the time of the Greeks there were well-known stories circulating about their exploits. The angels of Noah's time, these wandering stars, left their heavenly habitation and sinned, unlike the angels in heaven. God sent these sinful celestial angels to Tartarus, the deepest abyss of Hades. Abaddon was present long ago, but is not now because he is in the bottomless Abyss, and yet will come back physically to Earth's surface.

Humanity marvels at the great military and oration abilities of Abaddon, and then worships him after Abaddon miraculously heals from a deadly head injury. If Abaddon's humanlike "bio-shell" is an extension of his spirit, then it is logical that he could rebuild it since his life force is not dependent on his "bio-shell", as human souls are. Soon afterwards, his false prophet has humanity build a sentient "image" to help track down and kill any who will not worship Abaddon. If one were to write long ago of an artificial intelligence that infiltrates Earth's electronic devices to track down any who won't worship Abaddon, the account would read very similar to what is written in Revelation 13:14-15. A conscious intelligence like that of *The Terminator* or the *Age of Ultron* movies, but is at the command of an extremely intelligent, and very well-spoken, articulate, highly-advanced life form, Abaddon, who can regenerate his biologically body, and whose false prophet likely uses this sentient life to target believers for efficient termination. "Abaddon" literally means destroyer.

Assuming angels have large wings and use technologically illiterate 5,000-year-old human chariot technology might be part of the great lie to come. Will wayward angels deceive humanity by posing as evolutionarily advanced extraterrestrials with **great technological signs and wonders?** (Matthew 24:24; 2 Thessalonians 2:9-11)

Jesus was made a little lower than the angels when He came to Earth to reconcile the things on this rebellious Earth with the "other things" in the Universe, so it stands to reason that these "other things" have highly advanced technologies. Angels are much more advanced and greater in power and might than humans! Some angels are with God the Father who dwells in eternity, and some angels, evidently a vast assortment of them, dwell elsewhere in the Universe and very likely have been around much longer than humanity has, and with a vastly superior knowledge of exotic technologies that we know nothing about, which would be perceived by humanity as "alien tech". Humanity knows virtually nothing of quantum physics, gravitational waves, or dark energy. At one time humanity thought reproduction and healing were miracles, but now that we know more about it, we just think it is amazing. We know so little about the Universe, and nothing of its many types of inhabitants, but some day we will know much more of angelic exopsychology, physiology and physics than we do now. (John 5:4; 1 Corinthians 13:9-12)

Humanity has determined so far that everything God has made follows very precise scientific physical formulas and biological principles.

What if we tried to imagine Earth, and the only evidence we had were descriptions of two people walking on the sand, a sports car, and a mongoose. Could we then possibly imagine what Earth is like with all of its complexity of life, ecosystems, and the last several thousand years of modern human history? And so, we can't possibly know about the many other spirit-enabled sentient extraterrestrial life forms that God has created with their complexity and history and advanced technologies that may even go back for billions of years!

Did you know that Revelation 18 explained long ago that humanity will use a planetary merchandising system of cargo ships to transport vast quantities of costly and extravagant merchandise between countries? This ancient account lists what was costly merchandise at the time it was written a few thousand years ago— fine linen, purple, silk, ivory, food, **four-wheeled carriages**, and all manner of precious things that our souls lust for and desire (the Greek "rheda" is usually translated as chariots, but in ancient Rome a **rheda** was actually a **four-wheeled traveling carriage**, which was the closest thing there was to an **automobile** at the time it was written). This prophesy of Revelation 18 accurately depicts our current global economy with its costly merchandise shipments of designer clothes, food, extravagant smart phones, and the automobiles that our souls desire and lust for.

Revelation 18:23 also reveals that our vast global system will deceive the nations with her "pharmakeia". "Pharmakeia" is usually translated as "sorceries", but **pharmakeia** is the Greek root word for pharmacy, and implies that this deception involves both legal and illegal **drugs and medicines**. God made medicines for humanity to use, but, like most things, something God made for good can be twisted into evil. Drugs are now a global epidemic— Narcotic-based prescription drugs, although legal, are highly addictive with widespread abuse, as well as mood-altering drugs and recreational drugs. Could it be said that the nations are now deceived by illegal drugs, and by some legalized drugs? Revelation 18:23 accurately depicts our global dependency on drugs!

End time events are scary, but it is just a temporary transition for believers. After we die, our soul and spirit leave our physical body and go on to live with the Lord. God promised to someday give us new and vastly upgraded physical bodies. We may look back in pity on how inferior our old biological bodies were. The end game is not to do whatever it takes to stay alive, but to help others, and to reach others for Christ so they too can live on eternally!

Revelation 13:7, 17:6 and 20:4 makes it clear that many believers are to be slaughtered. The few who do survive to the end will be saved when their bodies are physically transformed into new ones as they still live! Imagine playing a computer game which gifts you just one life-force to help others. Whenever your life-force ends then you are guaranteed to level up for a vastly improved eternal body with amazing abilities in a much better place. This eternal guarantee of physically and spiritually leveling up after we die is what believers refer to as our "Hope".

Our Hope is our guarantee of our future life with Christ, not in stockpiling food and living in a hole in the ground to wait it out until Jesus returns. Expect to die, but try to die well. Jesus spoke of one who buried their talent in a hole and did nothing to invest it. Our treasure is our faith in Jesus, and our Hope is our future life with Christ and God. Investing our treasure and using our talent is being a witness for Jesus, so we shouldn't hide our talents! Yet, Revelation 13:10 warns us to be prudent, to not go into captivity or lead others into captivity, and if we fight with guns we will die with guns. Don't go into captivity and don't fight. This is God's advice— try and survive so we can help others.

Understanding the magnitude of God's predicament with Satan helps us understand our current state of eminent death, pain, sickness, wars and evil acts. The next chapter discusses what God has told us of this vast Universal conflict and how He is going to handle it. Humanity plays an important role in God's predestined plan of dealing with Satan's discord in the heavens, and we are now caught in the midst of this predestined heavenly war! So, if you feel like things are chaotic, well, they are! God, in His perfect time, will soon finish dealing with this Universal dissonance. We just need to be patient.

References used in composing this first chapter— Genesis 6:1-5, 18:1-19:22; 28:12-13; Exodus 34:29-30; Enoch 6; 2 Kings 2:12, 6:17; Isaiah 57:15; Daniel 7:13-14, 9:21, 10:6-19; Zechariah 5:1-6:7; Matthew 17:2, 22:30, 24:30, 26:64, 28:2; Mark 9:2, 13:26, 14:62; Luke 2:14, 21:25-27, 24:36; John 1:51, 3:16-17, 5:4, 20:26; Acts 1:9, 5:19, 8:39-40, 12:7 23:6, 24:15; Romans 8:20-24,38; I Corinthians 4:9, 6:3, 11:10; 15:38-53; Galatians 1:8; Ephesians 1:10, 20-23, 2:6, 3:10-14, 4:10, 6:12; Philippians 2:10; Colossians 1:15-20, 2:14-15; 1 Thessalonians 4:13-18; 1 Timothy 6:15-16; Hebrews 1:4, 2:5-9, 13:2,21; 1 Peter 3:19; 2 Peter 2:4-11; 1 John 3:2-3; Jude 6-15; Revelation 1:7, chapters 8 and 9, 12:12, 13:3-18, 17:8,18, 19:19, 21:2.

The Investigation

Litigation and investigating medical fraud have long been a part of my career. It occurred to me while writing *God and Humans* that by using a Hebrew and Greek Interlinear Dictionary, and reverse searches with the Strong's Concordance, I was instinctively using investigative techniques. Studying the source evidence of how words are used throughout the Old and New Testaments helps determine the truth and context in the original biblical text, and what these passages really indicate. If our viewpoint isn't aligned with the infallible Word of God then it isn't correct, regardless of how many others think it to be true. Only what is strongly supported by scripture should be embraced as a strongly held belief.

After growing up in a conservative Christian group in California, I gradually came to realize that some of what I had believed to be true in the depth of my soul was actually in error. Like peeling back the layers of an onion, I gradually came to realize how wrong I had been about some things that I had believed to be true to the core of my heart. Even my own conscience was flawed— some things that I had thought were wrong were actually acceptable, and some things that I had thought were right were actually quite wrong. This deeply disturbing, and painful, experience helped me realize how many of us have strongly felt beliefs that we haven't prayerfully studied and could be wrong about. We tend to assume that we already know the answers and must defend what we think, and we tend to discount any new information that conflicts with our opinions. Like most humans, Christians tend to unquestioningly believe what their respected friends believe, especially if they heard or read about it when they were young. So, in conducting my investigation, little consideration was given to the opinions of others. But, I *was* relieved to find that others had arrived at some of the same conclusions! (See the suggested reading at the end of this book.)

If something conflicts with your beliefs, I hope you will read the references and meditate and pray. We can be deceived and not even realize it. The Holy Spirit gently urges us towards the truth. (John 14:26)

With regard to matters requiring thought:
the less people know and understand about them, the more positively
they attempt to argue concerning them – Galileo Galilei

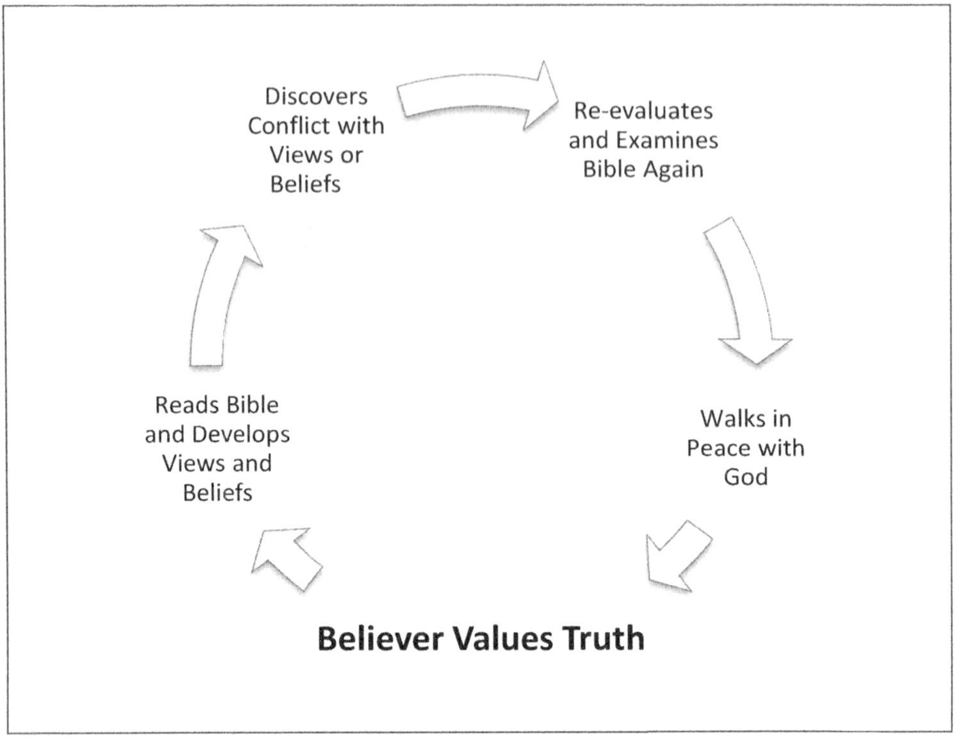

Test everything that is said. Hold on to what is good - 1 Thessalonians 5:19 NLT

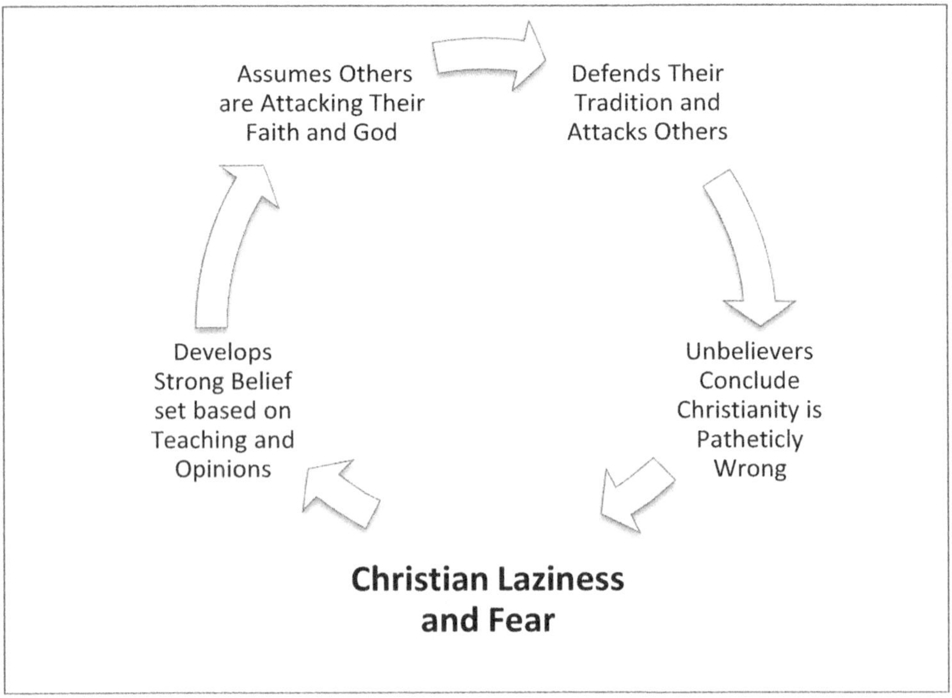

2

The Conflict in the Heavens

God's Dimension

We know so little of the Universe, of the vast heavens and of the spiritual dimension. We know so little of God's written requirements that govern the heavens, of what took place before humans, or even what is happening now with its masses of angels, principalities, powers, rulers of darkness, and spiritual hosts of wickedness. There are 100 million angels around God's throne alone. (Daniel 10:13,21; Zechariah 1:10-11; Ephesians 3:15, 6:12; Colossians 1:16, 2:15; 1 Peter 1:12; Revelation 5:11-14)

These extraterrestrial or extradimensional alien spectators are well aware of us living here on our little speck of earth, but faith is the only evidence that *we* have of *their* realms and abodes. By faith we can understand that past Eras were repaired by the invisible word of God.[1] (Hebrews 11:3) Similarly, scientists believe there is gravity and energy because of their evidence, even though they can't directly be seen and are not completely understood. (John 14:2)

Ezekiel got a fantastic glimpse into God's dimension of four amazing creatures with a human hand under each of their four wings. They fold two wings to cover their humanoid bodies as they fly in perfect wing-tip formation in the midst of bright fiery-amber whirlwind clouds of raging swirls of fire, with bursts of lightning flashing between and out of them. Their calf-like feet sparkle like gleaming bronze in fiery coals. These four creatures have four faces each; a man, a lion, an ox or cherubim, and an eagle. These faces may represent the four spiritual creations of God—Humans, Angels (cherubim/ox), Eagles (possibly spirit-enable birds of the heavens that fled in the Mesozoic Era), and Lions (Jesus is the Lion of the tribe of Judah). (Revelation 5:5)

The spirits of these creatures are in four massive transparent-yellow wheels, each with internal spinning wheels, which stay beside them as they fly perfectly straight, without veering, in any one of the four directions of their four faces. They are encompassed all around with eyes, and their thundering wings sound like a tremendous waterfall. They stand still and fold their wings when a voice comes from above.

High above them a tremendously vast crystal stretches out as a sea of glass, and high over this is the Lord God in human form who sits on a translucent deep-blue throne, surrounded by a translucent green rainbow. God's waist up is like a deep-yellow amber engulfed in fire, and His waist down radiates as a bright, fiery rainbow. In front of God's throne burn the seven lamps of fire with His seven Spirits, and around Him on thrones sit twenty-four elders, all wearing golden crowns. This is eternity, where God dwells. (Ezekiel 1; 10; Revelation 4; Isaiah 57:15)

The Trinity

Ezekiel states, from his own perspective, that God has human form, but it is we who have God's form. Why does God refer to Himself as "Us" and "Our" in Genesis 1, "Let Us make man in Our image, according to Our likeness"? Because the Triune God made humans. God the Father made everything through His Spirit and Jesus the Son.

John mentions all three in one passage: "By this we know that we abide in Him, and He in us, because He has given us of His **Spirit**. And we have seen and testify that the **Father** has sent the **Son** as Savior of the world. Whoever confesses that Jesus is the Son of God, God abides in him, and he in God. And we have known and believed the love that God has for us. God is love, and he who a bides in love abides in God, and God in him." - 1 John 4:13-16

God is one, and is also three: (1 x 1 x 1 = 1) 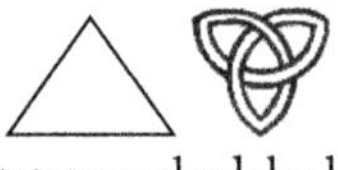

1) God the Father is love and dwells in unapproachable light.
2) God the Son is Jesus Christ the Lord and directly interacts with us in our physical dimension. Jesus lived in human form and died on the cross out of God's love for humanity. Jesus is like God's avatar for His physical creation, but He is also God, and is God's Son.
3) God the Holy Spirit is God's Spirit and gently motivates, comforts, teaches with whispers, is omnipresent, and is passively quenchable.

(John 1; 14:6-26; Ephesians 4:30; 1 Thessalonians 5:19; 1 Timothy 6:16; 2 Peter 1:21)

God the Father is known as the "Ancient of Days" or the "Lord of Hosts" in the Old Testament. God the Son was the "Angel of the Lord" and the "I AM" in the Old Testament, and was later born as Jesus in the New Testament, and is now known as the Lord Jesus Christ—

In Genesis 22:17, the Angel of the Lord tells Abraham, "I will bless you and multiply your descendants as the stars of the heaven and the sand on the seashore. Your descendant will take over and possess the gate of the hated adversary. Praised will be your descendant for all of Earth's nations."

In John 8, Jesus explains that He is from above, and that Abraham rejoiced in looking forward to His coming. The people ask Jesus, "You aren't even fifty years old. How can you say you have seen Abraham?"

Jesus answers, "Before Abraham was even born, I AM."

In Exodus 3:13-16, Moses asks the Angel of the Lord in the burning bush, "When I tell the children of Israel "The God of your fathers has sent me to you," and they ask, "What is His name?", what shall I say?"

God said to Moses, "I AM WHO I AM. Say to the children of Israel, I AM has sent me to you. The Lord God of your fathers, the God of Abraham, the God of Isaac, and the God of Jacob, has sent me to you. This is My name forever and is My memorial to all generations."

In Zechariah 1:12, the Angel of the Lord prays, "O Lord of Hosts, how much longer will You not have mercy on Jerusalem and the cities of Judah, which You were angry with for these seventy years?"

In John 17, Jesus prays, "Father, the hour has come. Glorify your Son so He can give glory back to you. You have given Him authority over everyone and He gives eternal life to each one you have given Him. This is the way to have eternal life—to know You, the only true God, and **Jesus Christ, the one You sent to Earth**. I brought glory to You here on Earth by completing the work You gave me to do. Now, Father, bring me into **the glory we shared before the world began**. May they have perfect unity so the world will know You sent me and that You love them as much as You love me. Father, I want these whom You have given Me to be with Me where I AM. Then they can see all the glory You gave me because You loved me even before the world began! Then Your love will be in them, and I will be in them."

(Genesis 16:8-13, 22:11-17, 31:11-13; Exodus 3:2-17; Judges 2:2, 13:9-23; Daniel 7:9-14; Zechariah 1; John 8:1-58)

Alice and E.T

A basic rule in writing movie scripts or books is that there has to be *something* normal people can relate to. Alice represents normality in the abnormal world of *Alice in Wonderland.* In the movie, *E.T. the Extra-Terrestrial*, the abnormal space creature E.T. is in a normal American suburb. But there is a lot going on in God's dimension that we don't yet know of, so it is difficult for God to communicate these things to us when we can't yet relate to them from our current perspective.

We are made in God's image and He loves us, so we can relate to God a little bit. We know a little of Satan. Like those living so long ago who were only told a brief synopsis of what occurred in Genesis, God has just told us what we need to know for now.

For Trekkies and *Doctor Who* fans, God may now have some analogies to draw upon to explain more of it. But the Bible was written thousands of years ago to people who knew nothing of *Doctor Who* or *Star Trek*. So, for them, and for us, God just informed us of what we humans need to know for now of His dimension and the Universe.

Overview of The Conflict in the Heavens

God and His Son Jesus create the sons of God and Lucifer. After this, Lucifer and these sons of God shout for joy when Earth is created.

Lucifer watches God make the first five major Eras. At some point, Lucifer is lifted in pride in wanting to be like God, and becomes a murderer. Lucifer loses access to the heavens and is cast to Earth and becomes Satan, or the devil, before Adam, also a son of God, is made. Satan assumes the body of the serpent and deceives Eve into influencing Adam to disobey God. Satan then obtains the power of death over humans with this sting of sin, gains control of Hades and Paradise, and becomes the reigning death god of planet Earth. Satan leverages this to regain access to the heavens, and can again present himself before God. So, when God placed Adam and Eve in the midst of this vast spiritual conflict, God already knew this would all happen, and already had a predestined plan to deal with Satan, even before He had created Earth!

God is still executing His plan when Satan enters the body of Judas to betray Jesus some two thousand years ago. Jesus dies to wrest this power of death from Satan, and then takes back with Him to heaven all of the souls of believers who had been held hostage in Paradise.

Jesus has already won and defeated Satan. We are just waiting for the clock to run out as the massive influx of believers is currently being added to God's realm. Satan is presently concealing himself as an angel of light, and his followers as ministers of righteousness.

Satan and his angels have a place they can stay at in the heavens, but Michael and his angels cast Satan out of the heavens and back to Earth. Then, at the end of the Great Tribulation, Satan is cast into the bottomless Abyss for one-thousand years while Jesus refurbishes Earth and peacefully rules during His one-thousand-year millennial reign of planet Earth.

Earth becomes a near-perfect social and technological utopia with peace, prosperity and long-life spans. At the end of His millennial reign Jesus lets Satan back out, and Satan quickly causes havoc on Earth. God then sends out a fire that consumes the planet and its atmosphere, burning everything except Jesus and His believers. After this, God permanently casts Satan into the lake of fire, along with everyone who has been held in Hades, and then casts Death and Hades into the lake of fire and destroys them. In the very end, God makes all things new.

There is more on this in the chapter, "The Spirit and Super Body."

~~~

Years ago, my brother-in-law moved across the country to live close to us for a fresh start, and soon married a friend of ours. They were happily married and very much in love. Six weeks after their wedding he was fatally injured in a construction accident, leaving his nineteen-year-old widow deciding when she should end his life support.

I was with her when he died, and it was a very impacting event of my life. Later, well-meaning people told her, "Well, it was his time to go and God wanted him home," and "God needed him so He called him home." This really disturbed her, and she told me in tears, "I needed him more than God needs him!" I told her people should be careful to not blame God for what Satan might be responsible for, and that I saw the hand of Satan in it (Paul, Silvanus, and Timothy wrote that they wanted to come, "<u>but Satan hindered us</u>"). I also saw human error. God doesn't stop all of the bad things from happening, but God *can* bring good from them, and His Spirit can comfort and teach us, and Jesus *has* made a way for us to be reunited. In time, my sister-in-law healed and happily married again.
(1 Thessalonians 2:18)
~~~

The Conflict in the Heavens

Jesus Christ the Lord

It was Jesus who gave life to **everything** that was ever created. He is from the beginning and was with God and He is God. (John 1:1-15)

"He is the image of the invisible God, the firstborn over all creation. For by Him all things were created that are in heaven and that are on earth, visible and invisible, whether thrones or dominions or principalities or powers. All things were created through Him and for Him. And He is before all things, and in Him all things consist."
- Colossians 1:15-18

"You are worthy, O Lord, to receive glory and honor and power; For You created all things, and by Your will they exist and were created." – Revelation 4:11

Jesus "was foreordained before the foundation of the world, but was manifest in these last times for you." - 1 Peter 1:20-21

Lucifer

Jesus created Lucifer as a son of the dawn, or a son of the morning. Lucifer shouted for joy when the foundation of Earth was laid—

"Where were you when I laid the foundations of the earth? Who laid its cornerstone, when the morning stars sang together, and all the sons of God shouted for joy?" (Job 38:4-7)

But afterwards Lucifer was lifted in pride, wanting to be like God—

"How you are fallen from heaven, O Lucifer, son of the morning! How you are cut down to the ground, you who weakened the nations! For you have said in your heart: 'I will ascend into heaven, I will exalt my throne above the stars of God; I will also sit on the mount of the congregation. On the farthest sides of the north; I will ascend above the heights of the clouds; I will be like the Most High.' Yet you shall be brought down to Sheol, to the lowest depths of the Pit. "Those who see you will gaze at you, and consider you, saying: 'Is this the man who made the earth tremble, who shook kingdoms, who made the world as a wilderness and destroyed its cities, Who did not open the house of his prisoners?'" - Isaiah 14:12-17 (Luke 10:18-19)

Did Lucifer shake kingdoms, make the world a wilderness, destroy its cities, and not open the house of his prisoners during the Mesozoic Era? Regardless of when or where it was, Satan fell from heaven because of it.

God revealed more of this to Ezekiel. The king of Tyre had supplied the gold and cedar for the temple that Solomon built, and he is an analogy of Satan in this account. Ezekiel's vision is of Satan's past—

"Thus says the Lord God: 'You were the seal of perfection, full of wisdom and perfect in beauty. You were in Eden, the garden of God; every precious stone was your covering: The sardius, topaz, and diamond, beryl, onyx, and jasper, sapphire, turquoise, and emerald with gold. The workmanship of your timbrels and pipes was prepared for you on the day you were created. You were the anointed cherub who covers; I established you; you were on the holy mountain of God; you walked back and forth in the midst of fiery stones. You were perfect in your ways from the day you were created, till iniquity was found in you.

By the abundance of your trading you became filled with violence within, and you sinned; Therefore I cast you as a profane thing out of the mountain of God; And I destroyed you, O covering cherub, from the midst of the fiery stones.'" - Ezekiel 28:11-16

Was Satan filled with violence during a fantastic Middle-Earth period of the Mesozoic Era? When God commanded the angel sons of the morning, is Satan the dawn that God caused to know his place by casting him out of heaven, when the wicked were shaken out of the extremities of the earth and their light was withheld? (Job 38:6-7, 12-15) Is Jeremiah 4:23-27 referring to the dark pre-human time when Earth became without form, and void, when the birds of the heavens fled, its cities were broken down, and Earth's atmosphere became so black it blocked all of the sun and star light? Was Lucifer involved in Earth's trembling destruction per Haggai 2:6? Could these Mesozoic Era inhabitants be the dead of the sea that are given up in Revelation 20:13? Suggesting some passages indicate that there may have been Mesozoic Era cities is about as crazy as Simon Episcupius teaching in the 1600's that he thought Genesis 1:2 indicates that there was an Era prior to ours. (Discussed in my book *God and Humans*)

Jesus said Satan was a murderer from the beginning and there is no truth in him, and when Satan speaks a lie he speaks from his own resources, for he is a liar and the father of lies.

Beelzebub, which means lord of the flies, or the dung god, rules over demons, but the origin of demons is not revealed. Satan might be Beelzebub, or may just have more authority than Beelzebub on Earth. Demons favor the sea, are associated with birds, and detest the bottomless Abyss that Jesus can command demons to go into.

Satan has drawn a massive one third of the angels to him, and at some point they fight Michael the archangel and other angels in a war of the heavens.
(Matthew 8:28-32, 12:24; Luke 8:31, 10:18; John 8:44; Revelation 12:3-12, 18:2)

God's Master Plan

God the Father knows the beginning from the end, since He inhabits eternity. No one can find out what God will do in the future. Only God from eternity knows all that will ever happen. (Exodus 3:14; Isaiah 57:15-18; Ecclesiastes 3:11; Colossians 1:16-17; Acts 15:18; Hebrews 13:8)

Before Adam and Eve, and all the way to our present time, there has been, and still is, this broader problem in God's realm involving Satan. God wants companionship and relationships with others. God is love, but God also wants peace, justice and order, and needs to deal with Satan. God creates humans in His image in Genesis 1, and they proliferate. It doesn't say if Satan had accused these humans of being blessed with copious amounts of self-control,[2] but when they need advanced abilities to sustain their dense populations, God makes Adam, a son of God.

God's plan to deal with Satan began before He made Adam the son of God (Luke 3:38) with the capability to choose to obey His instructions. God knew before they even took place of all the bad things that transpire in our world. He knew Adam would disobey and all of humanity would be held under Satan's power of death with this sting of sin. God already knew Satan would leverage Adam's sin to set up his own throne as the god of planet Earth with this power of death and would again gain access to heaven. (2 Corinthians 4:3-5; Hebrews 2:14) God already knew of all the problems that Adam's sin would cause. God hates sin because it harms others, but especially because sin separates us from Him.

So, way before Adam was made to be God's son, or Earth was even created, God had foreordained Jesus to overcome Satan, spiritual and physical death, and sin. Sin is not *a* problem: it is *the* problem. To end sin is to ultimately end all human suffering. That is why, billions of years ago, God pre-planned, or predestined, His son Jesus to be a sinless genetic descendent of His son Adam, and to pay the price of death for Adam's sin that will cause death and sin to spread to all of humanity. God already knows the angels will minister to these humans. (Genesis 3:15; Romans 5:12-17, 8:29; Hebrews 1:13-14; Peter 1:20-21; Revelation 5)

God predetermined a way for us to be reconciled with Him, so all we have to do is believe in Jesus to protect our souls from judgment. God's foreordaining of Jesus resolves the problem of sin for both on Earth and in the heavens. God knew, long before He created Earth, that all believers will someday live with Him on a new, renovated earth.

All believers will have already chosen to serve God, and to reject themselves and Satan, by believing in God during their life here on Earth. You see, there is still this problem with other angels later rebelling against God, as Lucifer and his angles have. So, God will have humans, in their new, vastly superior physical bodies, judge the angels in the future. This puts a fix in place for dealing with future fallen angels.

> "Do you not know that we will judge angels? How much more, things that pertain to this life?" - 1 Corinthians 6:3

God starts executing His predestined plan. God knows the future because He dwells in eternity. He made time, and time consists in Him.

"Even from everlasting to everlasting, You are God." - Psalms 90:2

Adam's Disobedience

Genesis seems simple when it is understood from Adam's limited perspective, but we now know that Lucifer was cast to Earth, entered or assumed the body of the cunning sentient serpent, and deceived Eve into influencing Adam to disobey God. (Revelation 12:3-9, 20:2)

Adam or Eve may have added in the additional "nor shall you touch it" to keep from being tempted by the fruit, since God said not to eat from the tree of the knowledge of good and evil, but said nothing of just touching it. This enlightenment of good and evil that Adam and Eve received went far beyond a mere understanding of consequences. After being deceived, desiring the fruit and disobeying God, they then saw good and evil from a supernatural perspective. Adam and Eve's eyes were opened and they received the perspective of clothed angels from the heavenly realm. No longer naive, they suddenly realized they were naked. This was a supernatural enlightenment of good and evil.

So, when God gives His son Adam a direct order to not eat from this tree, God already knows Adam will disobey and Satan will gain some control over Earth with the power of death with this sting of sin, that Jesus will descend from Adam and overcome this power of death, and for thousands of years billions of humans will need to decide if they should accept or decline God's gift of eternal life, and that Satan and his angels will be destroyed in the end, along with Death and Hades—

"Now the serpent was more cunning than any beast of the field which the Lord God had made. And he said to the woman, 'Has God indeed said, 'You shall not eat of every tree of the garden'?'

And the woman said to the serpent, 'We may eat the fruit of the trees of the garden; but of the fruit of the tree which is in the midst of the garden, God has said, 'You shall not eat it, nor shall you touch it, lest you die.''

Then the serpent said to the woman, 'You will not surely die. For God knows that in the day you eat of it your eyes will be opened, and you will be like God, knowing good and evil.'

So when the woman saw that the tree was good for food, that it was pleasant to the eyes, and a tree desirable to make one wise, she took of its fruit and ate. She also gave to her husband with her, and he ate. Then the eyes of both of them were opened, and they knew that they were naked; and they sewed fig leaves together and made themselves coverings. And they heard the sound of the Lord God walking in the garden in the cool of the day, and Adam and his wife hid themselves from the presence of the Lord God among the trees of the garden.

Then the Lord God called to Adam and said to him, 'Where are you?'

So he said, 'I heard Your voice in the garden, and I was afraid because I was naked; and I hid myself.'

And He said, 'Who told you that you were naked? Have you eaten from the tree of which I commanded you that you should not eat?'

Then the man said, 'The woman whom You gave to be with me, she gave me of the tree, and I ate.'

And the Lord God said to the woman, 'What is this you have done?'

The woman said, 'The serpent deceived me, and I ate.'"

- Genesis 3:1-13

Satan's Rule

Satan is the ancient serpent who deceives humanity, the great fiery red dragon and wonder in heaven who was punished and cast out of heaven to Earth with his followers, who gained control over Hades with the power of death to become the reigning death god of planet Earth. (Romans 5:17; 2 Corinthians 4:4; Hebrews 2:14; Revelation 12:3-9)

This was all due to Adam's disobedience as a son of God. Satan is now the prince of the power of the air, the spirit that now works in the disobedient, and the whole world lies under his sway. Satan gained the power of physical death with this sting of sin, and the humans are now vulnerable, and their souls are in peril. (Matthew 4:9; 1 Corinthians 15:55-56; 2 Corinthians 4:4; Ephesians 2:2; Hebrews 2:14; I John 5:19; Jude 9; Revelation 12:9)

The fruit from the tree of good and evil wasn't magical, but symbolic. Their punishment came after they ate the fruit. It is like pushing the symbolic red button for launching a nuclear arsenal, and then wondering later how that little red button could be so powerful to have caused all of that. Sin was released and evil spread throughout the entire human race. Humans, made in God's image, are now living and relating to one another with this supernatural enlightenment of knowing good and evil. Because of Adam's disobedience to God, sin entered the entire world, and the dominion of death spread to all men, reigning even over those who had not sinned like Adam had. Adam is a type of Jesus, who later ended what Adam had started. (Romans 5:12-17)

Under the influence of Satan and his angels, and their newly found knowledge of good and evil, humanity rejected God and sinned.

It isn't relevant if we are related to Adam,[3] but that we live on planet Earth where Satan is a god and gained power over our human souls with this power of death. Each human spirit is given by God, not from any genetic relationship to Adam, so the spirits of genetically altered humans are under sin also. We now know good and evil after being born as human, and are now accountable, as the angels are. (Ecclesiastes 12:7)

Lucifer, who had rebelled against God and was cast out of heaven, now, as the god of Earth, again has access to God's realm. Satan, as he did with Job and Joshua, is now incessantly accusing people of things to God in heaven, and demanding justice upon us. (Revelation 12:10) God is planning for the long term, but it is His divine will (Luke 11:2) that there is now discord in His dimension, and in the heavens, at this time—

Satan opposes and accuses Job—

"Now there was a day when the sons of God came to present themselves before the Lord, and Satan also came among them.

And the Lord said to Satan, "From where do you come?"

So Satan answered the Lord and said, "From going to and fro on the earth, and from walking back and forth on it."

Then the Lord said to Satan, "Have you considered My servant Job, that there is none like him on the earth, a blameless and upright man, one who fears God and shuns evil?"

So Satan answered the Lord and said, "Does Job fear God for nothing? Have You not made a hedge around him, around his household, and around all that he has on every side? You have blessed the work of his hands, and his possessions have increased in the land. But now, stretch out Your hand and touch all that he has, and he will surely curse You to Your face!"

And the Lord said to Satan, "Behold, all that he has is in your power; only do not lay a hand on his person."

So Satan went out from the presence of the Lord." – Job 1:6-12

Satan opposes and accuses Joshua—

"Then he showed me Joshua the high priest standing before the Angel of the Lord, and Satan standing at his right hand to oppose him.

And the Lord said to Satan, 'The Lord rebuke you, Satan! The Lord who has chosen Jerusalem rebuke you! Is this not a brand plucked from the fire?'

Now Joshua was clothed with filthy garments, and was standing before the Angel. Then He answered and spoke to those who stood before Him, saying, 'Take away the filthy garments from him.' And to him He said, 'See, I have removed your iniquity from you, and I will clothe you with rich robes.'" – Zechariah 3:1-4

In the past the Lord solicitated input from those in His realm; they stood by Him and offered their advice; but in the end, thrones are set up for twenty-four elders, and 100 million angels will proclaim Him worthy and minister to Him. (1 Kings 22:19-23; Daniel 7:9-10; Revelation 4-5)

Free will, though it makes evil possible, is also the only thing that makes possible any love or goodness or joy worth having
– C.S. Lewis, Mere Christianity

The Savior

All of us, even high priests, need help, a redeemer, since Satan is now accusing believers to God all day and night, or as it happened to Peter, is asking to sift us as wheat. (Luke 22:31-32; Revelation 12:10)

It is sin that keeps us from having a social utopia, and why all forms of human government, and space exploration, will have failure. Sin causes wars, crimes, social problems, selfish ambitions and injustices. But more importantly, sin separates us from God. Sin is not *a* problem, but it is *the* problem that troubles humanity everywhere. To end sin is to end human suffering. So, Jesus of Nazareth comes, God in a human body, and with power and the Holy Spirit, He helps others, healing all who are oppressed by Satan. (Acts 10:38)

In C.S. Lewis's book, *The Lion, The Witch, and The Wardrobe*, the boy Edmund sins under the trickery of the White Witch. The White Witch confronts the Lion and successfully argues that Edmund must die. But the Lion pays for Edmund's wrongdoing with his own life instead. However, since the Lion hadn't sinned, he comes back to life. The reason the Lion can pay for Edmund's wrongdoing and come back to life is because of a "deeper magic." The lion represents Jesus, who paid the price for our sins. (C.S. Lewis was led to the Lord by his friend, J.R.R. Tolkien, author of *The Hobbit* and *The Lord of the Rings*, who was also a believer in Jesus.[4])

The Lord, known on Earth as Jesus, knew, of course, of this "deeper magic". Satan had taken advantage of an existing law or requirement over humans because of Adam's sin. But Jesus bailed us out of that law and requirement by dying on the cross, so now all we need to do is ask forgiveness and believe in God to get God's gift of grace, a free pass to be with God and live forever physically in a new body.

Since Jesus had knowledge of good and evil, and was sinless, He conquered Satan and gained control over death—

"Because God's children are human beings– made of flesh and blood– the Son also became flesh and blood. For only as a human being could he die, and only by dying could he break the power of the devil, who had the power of death. Only in this way could he set free all who have lived their lives as slaves to the fear of dying."
– Hebrews 2:14-15 NLT

"But after Jesus died on the cross, He wiped out the handwriting of requirements that was against us, and disarmed it, nailing it to the cross. Then, 'having disarmed principalities and powers, He made a public spectacle of them, triumphing over them in it.'" - Colossians 2:13-15

After Jesus did all of this for us, He laid His right hand on John and said, "Do not be afraid; I AM the First and the Last. I AM He who lives, and was dead, and behold, I AM alive forevermore. Amen. And I have the keys of Hades and of Death." - Revelation 1:17-19

It's like Captain America bursting into a well-fortified stronghold, disarming the bad guys, rescuing the prisoners, and then triumphantly leading them out while holding the very weapons of death that were used to imprison them. Or a bit like Neo's control of the Matrix after his resurrection that rendered the Agents powerless.

Jesus made a public spectacle of them, triumphing over them in it. Since Jesus now holds the keys of Hades and Death, the gates of Hades, and Death, no longer spiritually prevails against believers, so our souls and spirits now skip Paradise and go directly to be with the Lord. (Matthew 16:18, 2 Corinthians 5:8)

We are not fighting against humans, but against spiritual crowds of wickedness in the heavenly places– against Satan and his angels, against the principalities, powers, and against those who rule the darkness of our time. (Ephesians 6:12)

Jesus, who is God, is also a direct genetic descendant of God's son Adam. Jesus lived a sinless human life and died to pay for all the sins of humanity. So now, any human who repents of their sin and gives up their own will, who accepts and submits themselves to God's will, is freely given God's Holy Spirit and the gift of eternal life by believing, and their names are not blotted out of God's Book of Life. Jesus paid the full price for Adam's original sin that afflicted all of humanity, and He will soon completely destroy Satan, Hades *and* Death.

Believers in Jesus can now say, "I am persuaded that neither death nor life, nor angels nor principalities nor powers, nor things present nor things to come, nor height nor depth, nor any other created thing, shall be able to separate us from the love of God which is in Christ Jesus our Lord." - Romans 8:38-39

All who dwell in the heavens now know of God's manifold wisdom and plan for believers— Paul said, "I was chosen to explain to everyone this mysterious plan that God, the Creator of all things, had kept secret from the beginning. God's purpose in all this was to use the church to display His wisdom in its rich variety to all the unseen rulers and authorities in the heavenly places. **This was His eternal plan**, which He carried out through Christ Jesus our Lord. Because of Christ and our faith in Him, we can now come boldly and confidently into God's presence." - Ephesians 3:9-12 NLT (Ephesians 4:7-10; Acts 10:45)

The invisible God made through Jesus everything that exists in the heavenly realms and Earth. Jesus is supreme over all creation, and existed before time or space or anything that was created. Everything was created for Jesus, and He holds all of creation together. Jesus made everything that we can see and everything we can't see, such as the thrones, kingdoms, mansions, rulers, and authorities in unseen worlds. God predetermined before Earth was even created that Jesus Christ the Lord would live on Earth as Jesus and die for our sins. God had Jesus purchase our freedom so He can forgive our sins and rescue us from the kingdom of darkness and transfer us into His Kingdom.
(Jeremiah 33:22; 2 Kings 22:13-28; John 14:2; Colossians 1:13-17; 1 Peter 1:20-21)

Satan's System of Hypercriticism

Satan is involved in drugs and immoralities, but also, in a truly more diabolical effective way, Satan subtly disguises himself as an angel of light, and his servants as ministers of righteousness. There are many wickedly hypocritical religious things they cozen humans to do in religion that are erroneously blamed on God. How more devilish of a way is there for Satan to fight against God, than to pose as God and influence people to do depictable things so that humanity blames God for them?[5]

Paul wrote, "I will continue doing what I have always done. This will undercut those who are looking for an opportunity to boast that their work is just like ours. These people are false apostles. They are deceitful workers who disguise themselves as apostles of Christ. But I am not surprised! Even Satan disguises himself as an angel of light. So it is no wonder that his servants also disguise themselves as servants of righteousness. In the end they will get the punishment their wicked deeds deserve." - 2 Corinthians 11:12-15 NLT

Satan, the god of this age, is blinding the minds of unbelievers to conceal the gospel so they can't see the light of the gospel that reveals the glory of Christ, who is the image of God. (2 Corinthians 4:3-4)

Jesus told the self-righteous poisonous snakes of His time, "If God were your Father, you would love Me because I came from God. I did not come for Myself, but God sent Me. Why can't you understand Me? Because you are not able to listen to My word. Your father is the devil, and you desire to do whatever he wants you to do. He was a murderer from the beginning, and does not stand in the truth because there is no truth in him. When he speaks a lie, he speaks from his own resources, for he is a liar and the father of it. But because I tell you the truth, you do not believe Me." (John 8:42-46; Jeremiah 12:2, 29:14, 38:24-26)

> Satan's masquerade as a religious angel of light, and his followers as religious apostles of Christ, are highly successful schemes for blinding the minds of unbelievers to Christ. Either of these two extremes may indicate that Satan or his apostles are influencing a church or religion—
> 1) Controlling legalism with misuse of scripture, requiring compliance with the leadership's doctrines, and devaluing of relationships.[6]
> 2) Broad permissiveness of conduct, and a vague reliance on scripture.

(Mark 8:15; Galatians 5:16-26; 2 Timothy 2:15-18; Titus 3; 1 Peter 1:14; 3 John 9-10)

Legalism is Satan's assault on God's wisdom and insight.[6]

Ignorance and stupidity give Satan his broad destructive power.

Actor Chris Pratt, of Guardians of the Galaxy and Jurassic World fame, while surrounded by large glorifying posters of himself in a pinnacling moment of his career during the MTV award of 2018, gave glory to God and pointed people towards Christ.[7] It was laced with comments I wouldn't want my kids to repeat, *but* it was appropriate for an MTV show, and MTV did not publicly object. Chris says he has a burden for young people and some may not like how he goes about it. (Luke 9:50) Chris said, "I would not be here...without my Lord and Savior Jesus Christ."[8] As Jesus said, if God is our Father, then we love Jesus.

Ironically, some religious folk took to the internet with accusations, insinuating Chris was an agent of Satan…— a tendency towards making judgmental accusations is Satan-like, not Christ-like. Jesus wants us to be impartial with others. To be partial is to be a judge with evil thoughts. Mercy triumphs over judgment. (Luke 9:49-50; James 2:1-14)

The Future

Michael is the great angel prince who watches over the Jewish people, and there are many angels in heaven who fight with him.

Satan, who is called the serpent of old, the great fiery red dragon, or the devil, has a place in the heavens that he can go to now with his wicked angels, the principalities and powers who rule in the darkness of our time. Satan is responsible for deceiving humanity. (Revelation 12:9)

In the future, during the Great Tribulation, a war breaks out in heaven when Michael and his angels fight against Satan and his angels. Near the end of this war, the place where Satan and his angels can go to in heaven is terminated, and Michael and his angels are able to cast Satan and his angels out of the heavens and down to Earth.

When Satan and his angels are finally cast out, a loud voice in heaven says, "Now salvation and strength and the kingdom of our God and the power of His Christ have come, because the accuser of our brethren, who accused them before our God day and night, has been cast down. They defeated Satan with the blood of the Lamb and the word of their testimony, and they did not love their lives to the death. **So be very glad, all you who dwell in the heavens**, but watch out, all you who dwell on the earth and are of the sea! For Satan is coming down to you in a great rage, because he knows his time is running out."

After Satan realizes he can't get back into the heavens, he and the angel Abaddon persecute Israel, because Jesus was Jewish, but the Jews take flight and hide. Satan knows the God of peace will crush him soon, and quickly causes millions of believers to be slaughtered. Jesus Christ soon comes back to Earth with His army in great clouds and defeats Satan. An angel then locks Satan up in the bottomless Abyss, and sets a seal on Satan so he can't deceive the nations for one thousand years.

The Lord then reigns for one thousand years on Earth. Near the end of the Lord's millennial reign, Satan is released from the bottomless Abyss and soon deceives all of the nations into attacking the Lord's holy city. Fire comes down from God and devours the attackers, and Satan is then cast into the lake of fire forever. All who rejected God will then be destroyed, and then finally Hades and Death are destroyed.

God will then descend in His incredible new city and dwell with Humans. There will be a new, fantastically beautiful Earth. (Daniel 9:27, 10:13,21, 12:1; 1 Corinthians 15:23-24; Jude 9; Revelation 9, 12:7-14; 20, 21)

The Importance of Jesus's Genealogy

God's foreordaining of Jesus to reconcile humanity with Him was predestined even before He created Earth. It is God's will that this be done. (Matthew 6:10; 1 Peter 1:20-21; Romans 3:27-28, 4:4-5, 5:1-2, 8:29-30, 10:9-13, 11:32; John 3:16; Revelation 3:20, 5:1-14)

This is why Genesis 5:1 states, "This is the book of the genealogy of Adam," and why Luke gives Mary's biological genealogy of Jesus, to make it clear to us that Jesus the Son of God is a sinless descendant of Adam the son of God. (Luke 3:38)

Matthew prophetically telescoped the royal genealogy of Jesus to give insight into Daniel's prophesy of Jesus, and symbolically provides six weeks of generations to point out that Jesus is the seventh generation, the Millennium Lord of the Sabbath.
(Matthew 1:11,17; Revelation 5; Daniel 9:24-25)

Adam's sin spread to all humans, and there is now a supernatural battle between Jesus and Satan for human souls, and by faith alone we know that Jesus satisfied these laws and written requirements in the spiritual realm to gain control of the souls in Paradise. (Colossians 2:13-14)

"When Adam sinned, sin entered the world. Adam's sin brought death, so death spread to everyone, for everyone sinned. For the sin of this one man, Adam, brought death to many.

But even greater is God's wonderful grace and his gift of forgiveness to many through this other man, Jesus Christ. And the result of God's gracious gift is very different from the result of that one man's sin. For Adam's sin led to condemnation, but God's free gift leads to our being made right with God, even though we are guilty of many sins.

For the sin of this one man, Adam, caused death to rule over many. But even greater is God's wonderful grace and his gift of righteousness, for all who receive it will live in triumph over sin and death through this one man, Jesus Christ.

Yes, Adam's one sin brings condemnation for everyone, but Christ's one act of righteousness brings a right relationship with God and new life for everyone. Because one person disobeyed God, many became sinners. But because one other person obeyed God, many will be made righteous. God's wonderful grace rules give us right standing with God, resulting in eternal life through Jesus Christ our Lord."
(Passages from Romans 5 NLV)

Accusations and Grace

Evil is hurting other people for enjoyment.

Evil is hurting innocent children and people to feel powerful.

Evil is having reckless regard for others and hating them.

Evil is deceiving naive people to hold all human souls hostage, and then relentlessly making accusations against them.

~~~

Good is a superhero stopping a crazed robot from killing a child.

Good is stopping a bus full of innocent people from going off of a bridge that a villain has pushed with a semi-truck.

Good is helping others, even at great risk and peril to ourselves.

Good is being kind and considerate in small ways.

Good is stopping self-righteous hypocrites from stoning a woman to death for committing adultery.

Good is taking care of a poor man and paying for his medical care with no expectation of any benefit or recognition for doing it.

Good is self-sacrificing, giving for the good of others.

Good is having mercy and forgiving others.

Good is loving people.

Good is Jesus dying on the cross for the sins of the world, and then freely offering love, salvation and eternal life to everyone.

Good is God's grace, His undeserved goodness on all.

Only God is truly good. (Matthew 5:16,19:17; John 3:16; Revelation 12:10)

Understanding this is a supernatural enlightenment

The Crucifixion by Gustave Doré
~~~

3

The Book of Life

If we knew all aspects of quantum physics and DNA knowledge, even more than as it is portrayed in science fiction movies, and even if human knowledge progressed to the point where we understood everything there is to know about the Universe, it still wouldn't measure up to knowing the foolish things of God, which are still wiser than all human wisdom.
(Isaiah 40:12-15, 26, 55:8-11; 1 Corinthians 1:25)

God conceptualized physics, and the Universe instantaneously obeys His decrees and materializes.[9] Some try to describe His physics with the pure language of mathematics, but it is still quite difficult for even the most intelligent human minds to comprehend a timeless eternity beyond God's precise laws of physics, beyond what mathematics describes.[10,11]

Considering how vastly intelligent God is, does He even care about us? How do we compare to God? God made the tiny microbes that crawl on our skin to defend us from harm before our immunity system can even react.[12] Our microbes outnumber our human cells, yet are so miniscule they account for only about 3% of our body mass. Do you love your little miniscule microbes that you need? God dwells in an unapproachable light that no one has ever seen, and God doesn't need us. (Acts 17:25)

Like the galaxy inside the marble dangling from the collar of the cat in *Men in Black*, we are insignificant. We are also corrupt— try not showering for a few months. But, in spite of all of this, we are made in God's image and He loves and values us very much.
(Genesis 1:26; John 3:16; 1 Corinthians 15:53; 1 Timothy 6:16)

We can live nice moral lives and help others the best we can. Some of the nicest people I know do not claim to be Christians, yet care for others at a deeper level than many legalistic Christians do.

However, in the very end, when we stand before God the Father, it doesn't matter how moral we were, or what good deeds we did, or how much we cared for others. He will just check for our names in the Lamb's Book of Life. Everyone's name was written in it since the foundation of the world. But some names are blotted out.

The names that are blotted out **are not** made known to God, and they are later cast into the lake of fire and die the second death. But if our name has *not* been blotted out, then in the end Jesus will confess our name to God the Father, and we become **eternally** known to God, from the beginning to the end, and are allowed in. Not getting our name blotted out of this Book of Life is the key for God letting us in. Can we control where our soul goes after our body expires?
(Daniel 12:1; Matthew 7:23, 25:12, 25:41; Revelation 3:5, 20:12, 15)

Is Heaven like the dimension described in *The Forgotten Door*, by Alexander Key?[13] The people of this other dimension read each other's minds, which is okay because they have no evil thoughts. They live peaceful lives in simple homes in a great climate with plenty of fruits and vegetables to eat. They create art, visit, and learn new things.

Intriguing glimpses of God's dimension, or perhaps the future heaven and earth, might also be seen in Akiane Kramarik's paintings. Her parents were atheists, yet she was led to God and became an amazingly talented artist at a young age. She painted her inspirations, or revelations, some of which depict God's dimension. Her parents later became believers because of her faith. Her *Prince of Peace* painting, completed when she was eight, is impacting.[14]

Other glimpses of God's dimension might be seen from the little boy's perspective in the book, *Heaven is for Real.* He almost died, and in the weeks afterwards described seeing things in Heaven matching biblical concepts that many adult Christians don't understand. His father wanted to know what Jesus looked like and showed the little boy paintings of Jesus. The little boy said Jesus didn't look like any of the traditional religious paintings of Jesus, until he saw Akiane's *The Prince of Peace* painting, and said, "Yes, that is exactly what Jesus looks like!"

The little boy told of Jesus teaching the deep things of God, of Satan coming to heaven at times with fighting and conflict, and of time passing both very quickly and very slowly in heaven.

God says one day is as a thousand years, and a thousand years as one day. So, time is simultaneously both incredibly slow and fast in God's dimension. That is a difficult concept to grasp, God dwelling in eternity, knowing all things from the beginning to the end.

God is eternal and never changes. When Moses wanted to know the Lord's name, God replied, "I AM WHO I AM. Thus you shall say to the children of Israel, 'I AM has sent me to you.'" (Exodus 3:14)

Jesus Christ is the same yesterday, today, and forever. (Hebrews 13:8)

Jesus created all visible and invisible things in Heaven and Earth; all of the angels, thrones, dominions, principalities and powers. They were all created for Him. Jesus is before all these things, and by Him they exist. (Colossians 1:16-17)

God is timeless and inhabits eternity. (Isaiah 57:15)

God is from everlasting to everlasting. (Psalms 90:2)

"He has made everything beautiful in its time. Also He has put eternity in their hearts, except that no one can find out the work that God does from beginning to end." - Ecclesiastes 3:11

"Known to God from eternity are all His works." - Acts 15:18

When a believer becomes known to God the Father, He has **eternally** known them, from eternity past to eternity future. But those who never believe are never known to God, so God will have **never** known them.

The Book of Life

Moses spoke of God's Book—

"Now it came to pass on the next day that Moses said to the people, 'You have committed a great sin. So now I will go up to the Lord; perhaps I can make atonement for your sin.' Then Moses returned to the Lord and said, 'Oh, these people have committed a great sin, and have made for themselves a god of gold! Yet now, if You will forgive their sin - but if not, I pray, blot me out of Your book which You have written.' And the Lord said to Moses, 'Whoever has sinned against Me, I will blot him out of My book.'" - Exodus 32:30-33

King David spoke of it—

"Your eyes saw my substance, being yet unformed.
And in Your book they all were written,
The days fashioned for me,
When as yet there were none of them." - Psalms 139:16

Paul and Timothy wrote of it—

"And I urge you also, true companion, help these women who labored with me in the gospel, with Clement also, and the rest of my fellow workers, whose names are in the Book of Life." - Philippians 4:3

The Apostle John wrote about this Book of Life in the future—

"He who overcomes shall be clothed in white garments, and I will not blot out his name from the Book of Life; but I will confess his name before My Father and before His angels." - Revelation 3:5

"Then I saw a great white throne and Him who sat on it, from whose face the earth and the heaven fled away. And there was found no place for them. And I saw the dead, small and great, standing before God, and books were opened. And another book was opened, which is the Book of Life. And the dead were judged according to their works, by the things which were written in the books. The sea gave up the dead who were in it, and Death and Hades delivered up the dead who were in them. And they were judged, each one according to his works. Then Death and Hades were cast into the lake of fire. This is the second death. And anyone not found written in the Book of Life was cast into the lake of fire." - Revelation 20:11-15

"But I saw no temple in it, for the Lord God Almighty and the Lamb are its temple. The city had no need of the sun or of the moon to shine in it, for the glory of God illuminated it. The Lamb is its light. And the nations of those who are saved shall walk in its light, and the kings of the earth bring their glory and honor into it. Its gates shall not be shut at all by day (there shall be no night there). And they shall bring the glory and the honor of the nations into it. But there shall by no means enter it anything that defiles, or causes an abomination or a lie, but only those who are written in the Lamb's Book of Life."
- Revelation 21:22-27

It is super important to do a very simple thing to prevent your name from being blotted out of the Book of Life. God has warned us that if our name is blotted out, we will be tormented in Hades, and then later, at the great white throne judgment, Jesus will say, "I never knew you, depart from me", and we will be cast into the lake of fire and either burn forever or just cease to exist, a second-death worst-case scenario with no possible escape. (Matthew 7:23, 25:41; Revelation 20:11-12)

God gave us life and loves and wants us, but He doesn't need people who lived a good life. What right have we to force God to transport our soul and spirit into His kingdom and demand from Him a new physical body on our terms? God put that in us, to expect He should be kind and fair, but doing good doesn't get us in on His free gift pass and angelic transportation into God's Kingdom. He wants us to feel our way towards Him and find Him, although He is not far from any of us, for in Him we live and move and exist. We are God's offspring, and He overlooked our past ignorance, but has let us know He just wants those who believe in Him. Fortunately, He made it incredibly easy to get on His side of a terrible conflict between powerful spiritual beings. (Acts 17:18-32; Romans 3:21-26)

At the end of our Era terrible things happen. Some wealthy hedge fund managers sense this "big event" is coming, and their wealth and food will be targeted by others. How are they to keep order, have explosive collars around their guard's necks to force obedience?[15] Peace will be taken from humanity with mass civil unrest and violence, famine, death, and vast fires raging over a third of Earth. God has this tribulation at the end of our Era in hopes some will repent. (Revelation 6-21)

God *could* instantly destroy Earth's inhabitants, but allows the Great Tribulation to mercifully allow even those who repent at the very end to be eternally ushered into His new earth. (Revelation 9:21, 16:11, 11:13, 16:9) "As I live," says the Lord God, "I have no pleasure in the death of the wicked, but that the wicked turn from his way and live. Turn, turn from your evil ways! For why should you die?" (Ezekiel 33:11)

God knows some terrible thing that happens in the vapor flash of our life is worth it if we end up living eternally with Him. Remember having antiseptic medicine put on your cut when you were a kid, or having someone get a splinter out? It stings and is painful, but it feels better after a little bit and heals up and is soon forgotten. We stand on the brink of our own eternal destiny, and any horrible event at the end of our brief life is quite insignificant if it leads to our doing that one thing to prevent our name from being blotted out of God's Book of Life.

You see, God is still trying to reach us humans, even to the very end, because He genuinely loves us. It's like God having Jonah warn the Assyrians of Nineveh that He will overthrow them in forty days in an inevitable destruction, but they turn from their evil ways, so God relents and doesn't bring this disaster upon them. Jonah was very angry that God did not destroy the Assyrians because they had been extremely cruel to Jonah's people. Jonah said he had expected "God would relent, since God is gracious and merciful, slow to anger and abundant in loving kindness, One who relents from doing harm." (Jonah 4:2)

God wants a relationship with each of us, so He gives us the ability to choose and obey Him, which also allows for disobedience. All of our problems arise from knowing good and evil and being under this condition of sin, which separates us from God. So, God, who is love, even before creating Earth, predetermines the way of salvation so we don't have to be expelled from Him. (1 Peter 1:20-21)

I think the wise woman of Tekoah said it best. The back story on this wise woman is that King David, although he is a man after God's own heart and wrote Psalms and kills Goliath, he also commits adultery with his neighbor, Bathsheba. After David finds out Bathsheba got pregnant, he kills off her loyal husband in a very sneaky way. God, of course, saw all of this and was displeased, and let David know he and his family will have a lot of conflict because of it.

This conflict begins with David's first-born son Amnon becoming consumed with love and yearning to have sex with his half-sister Tamar, but he can't because it's improper. A crafty man then explains to Amnon how he can create this situation so Amnon can finally have sex with her. It works, except Tamar tries to talk her way out of it, so Amnon just rapes her. Right afterwards, Amnon hates her so much that the hate with which he hates her is greater than the love with which he had loved her.

Then, Amnon tells her, "Arise, be gone!"

So she says to him, "No, indeed! This evil of sending me away is worse than the other that you did to me."

But Amnon won't listen, calls in his servant, and says, "Here! Put this woman out, away from me, and bolt the door behind her!"

Tamar's brother, Absalom, is infuriated when he finds out. But he carefully conceals his fury, and secretly plots for two years before finally killing Amnon for raping his sister Tamar.

King David punishes his son Absalom for this by banning him from the royal city. Three years later, King David's general, Joab, sees that David is really missing Absalom, so Joab enlists this wise woman from Tekoah to help persuade King David to let Absalom come back. This wise woman makes up a story about herself and her two sons to get King David to understand her situation, and then turns the table on David and says the same situation applies to him and Absalom. At the end of these sordid events, the wise woman of Tekoah makes this very profound statement—

"For we all die, and our lives become like water spilled out on the ground, which cannot be gathered up again. But **God does not take life away; instead He devises ways to bring us back when we become separated from Him.**" (2 Samuel 13:1-14:14)

God devised a way to bring His banished children back to Him. This is why Jesus, God in human form, died for us. The fact is, we will get our names blotted out of the Book of Life unless we do something. God wants us to do that something. He *wants* us with Him.

The human minds that are controlled by Satan are so passively dependent on their bondage that they fight to remain enslaved, and they are so blinded by Satan's deception that they often can't see this liberty that God is offering them. God wants us to turn from Satan's darkness to His pure light and receive forgiveness of sins, to be made holy by faith in Jesus. (John 8:34; Acts 7:57, 26:18; 2 Corinthians 4:4; Galatians 4:7)

Believing and calling out to God for help and forgiveness prevents our name from being blotted out of God's Book of Life, and we become eternally known to the eternal God Almighty! The Lord says, "Repent, and turn from all your wrongdoing so your sins won't be your ruin. Throw away all the wrong things you have done against Me and get a new heart and a new spirit. Why should you die? I have no pleasure in the death of one who dies. Therefore turn and live!" (Ezekiel 18 paraphrased)

Many of us live quiet peaceful lives and haven't done anything close to what David, Amnon or Absalom did. But we may have done a few things on God's list of six distinct sins that He hates, and the seventh is one that God really hates—

1) Looking proud
2) Telling lies
3) Killing innocent people
4) Thinking of wicked plans to harm others
5) Being quick to do wrong things
6) Telling lies to pervert justice and harm others
7) Doing or saying things to damage the good relationships of others.
(Proverbs 6:16-19)

Even our own conscience can deceive us in what we believe to be true to the core of our being. The Lord searches the innermost intentions of our spirit. We will never measure up to God's standard.
(Jeremiah 17:9-10; Romans 4:5; I Corinthians 10:29; Colossians 2:16)

God will check in with Jesus's Book of Life at the end, and if your name is written in it then Jesus will confess your name before God and His angels, and you receive His free gift pass for eternal life.
(Matthew 7:23,25:41; Revelation 3:5)

It is believing in God and calling on the Lord and putting our faith in Him that prevents our name from being blotted out—

Jesus said, "For God so loved the world that He gave His only begotten Son, that whoever believes in Him should not perish but have everlasting life. For God did not send His Son into the world to condemn the world, but that the world through Him might be saved. He who believes in Him is not condemned; but he who does not believe is condemned already, because he has not believed in the name of the only begotten Son of God. And this is the condemnation, that the light has come into the world, and men loved darkness rather than light, because their deeds were evil. For everyone practicing evil hates the light and does not come to the light, lest his deeds should be exposed. But he who does the truth comes to the light, that his deeds may be clearly seen, that they have been done in God." - John 3:16-21

"But, beloved, do not forget this one thing, that with the Lord one day is as a thousand years, and a thousand years as one day. The Lord is not slack concerning His promise, as some count slackness, but is longsuffering toward us, not willing that any should perish but that all should come to repentance." - 2 Peter 3:8-9

"We are made right with God by placing our faith in Jesus Christ. And this is true for everyone who believes, no matter who we are. For everyone has sinned; we all fall short of God's glorious standard. Yet God, with undeserved kindness, declares that we are righteous. He did this through Christ Jesus when He freed us from the penalty for our sins. For God presented Jesus as the sacrifice for sin.

People are made right with God when they believe that Jesus sacrificed His life, shedding his blood. This sacrifice shows that God was being fair when He held back and did not punish those who sinned in times past, for He was looking ahead and including them in what He would do in this present time. God did this to demonstrate His righteousness, for He Himself is fair and just, and He declares sinners to be right in His sight when they believe in Jesus."
- Romans 3:22-26 NLT

"If you confess with your mouth the Lord Jesus and believe in your heart that God has raised Him from the dead, you will be saved. For **with the heart one believes** unto righteousness, and with the mouth confession is made unto salvation. For the Scripture says, 'Whoever believes on Him will not be put to shame.' For there is no distinction between Jew and Greek, for the same Lord over all is rich to all who call upon Him. For 'whoever calls on the name of the Lord shall be saved.'"
- Romans 10:9-13

"Therefore He is also able to save to the uttermost those who come to God through Him, since He always lives to make intercession for them." - Hebrews 7:25

'Come now, and let us reason together,' says the Lord,
'Though your sins are like scarlet, they shall be as white as snow'
- Isaiah 1:18

"For I am persuaded that neither death nor life, nor angels nor principalities nor powers, nor things present nor things to come, nor height nor depth, nor any other created thing, shall be able to separate us from the love of God which is in Christ Jesus our Lord."
- Romans 8:38-39

What was I comparing this Universe with when I called it unjust?
— C.S. Lewis

God says there *are* no atheists— "They know the truth about God because He has made it obvious to them. For ever since the world was created, people have seen the earth and sky. Through everything God made, they can clearly see His invisible qualities— His eternal power and divine nature. So they have no excuse for not knowing God. Yes, they knew God, but they wouldn't worship Him as God or even give Him thanks. And they began to think up foolish ideas of what God was like. As a result, their minds became dark and confused. Claiming to be wise, they instead became utter fools.

You may think you can condemn such people, but you are just as bad, and you have no excuse! Since you judge others for doing these things, why do you think you can avoid God's judgment when you do the same things? Don't you see how wonderfully kind, tolerant, and patient God is with you? Does this mean nothing to you? Can't you see that His kindness is intended to turn you from your sin?"
(Romans 1:20-22, 2:1,3-4 NLT)

If the thing loved is base, the lover becomes base
- Leonardo Da Vinci

Jesus said, "I AM the door. If anyone enters by Me, he will be saved, and will go in and out and find pasture. The thief does not come except to steal, and to kill, and to destroy. I have come that they may have life, and that they may have it more abundantly." – John 10:9-1

Merely realizing Jesus is the door is much different than opening that door and walking in. Jesus doesn't force His way into our heart, we must decide to open that door. Jesus offers all humans the gift of spiritual rebirth into a new spiritual being who is supernaturally instilled with God's Holy Spirit. All believers can, and should, look forward to physically living forever with God in our new physical, refabricated superbodies on a new physical earth. (Isaiah 65:17; John 3:16; 2 Peter 3:13)

Is it not My ways which are fair
and your ways which are not fair?
- The Lord
(Ezekiel 18:29)

The Abundant Life for Believers

Believers who used their talents to serve God and others will have their life's work of gold and silver purified in God's judgment—

"Now if anyone builds on this foundation with gold, silver, precious stones, wood, hay, straw, each one's work will become clear; for the Day will declare it, because it will be revealed by fire; and the fire will test each one's work, of what sort it is. If anyone's work which he has built on it endures, he will receive a reward. If anyone's work is burned, he will suffer loss; but he himself will be saved, yet so as through fire."
- 1 Corinthians 3:12-15

Selfish believers will have little to show for their life's work of wood, hay and straw after it is consumed in the heat of God's judgment of their actions. They will be like a wheat plant bearing few kernels with little benefit to others, but their soul and spirit will be saved since their name wasn't blotted out of God's Book of Life.

For an abundant life in Christ

Jesus is our rock and the protector of our faith. We know we will live eternally with God and can live confidently since we have been spiritually reborn as children of God. So, we should diligently build on our foundational faith and the peace we have made with God, and grow in Christ by availing ourselves of God's undeserved goodness.

First, **desire moral excellence** to break the deliberate bad habits and lustful desires of our body and mind by learning to walk carefully as a child of God in His light. We can now take these sins that burden and so easily trip us up to God, since we now have the undeserved privilege of God's love pouring into our hearts.

Reverence God's word and be glad to learn His truth and wisdom and walk in it. God's insights protect and weaponize us to defeat Satan's subtle strategies— we aren't fighting humans, but against evil rulers and authorities from the unseen world who are mighty powers in this dark world, and against evil spirits in the heavenly places.

Find mature spiritual mentors who can encourage us, and meditate on the huge crowd of witnesses throughout the Bible as examples of faith for how we should conduct our own lives and what decisions we should make. Open yourself up to God's Spirit and His leading.

We can **run with endurance** the race God has set before us by keeping our eyes on Jesus, our champion, who endured such a brutal death and public humiliation from evil self-righteous people, but is now seated beside God. So, don't get tired and give up, since you have not yet given your life in your struggle against sin! We *can* reap the harvest of living right when we are trained in His ways. How often should we pray? Always. We can't overdo praying. God wants to hear from us all the time. He always enjoys hearing from us and wants to help!

After we **learn to live self-controlled moral lives**, we need to take care to not be lifted in pride, because when the trials in life come, and they will, our pride will bring us down. But, Jesus is there to lift us up on our feet as many times as we ask. We should really be glad when our problems come, because it is through patiently enduring them that we grow in our faith and our character grows stronger. The burden Jesus places on us is light and easy. Our soul will rest when we make our requests known to Him, learn to trust Him by carefully reading God's word, and walk in harmony with Him. We can not only *stand* with God's help in the times of our trials, but *remain* standing.

We should **have patience during our trials** because it is the very problems in our lives that reveal the weaknesses in our spirit, the hidden faults of our heart that God can help us with to develop our strength of character. These trials are the times that God is healing our spirit, so don't make light of the Lord's divine discipline, or give up as He corrects, because He does this out of His love in wanting us to share in His holiness since we are His children. This shows He really does love and care for us! He is the Father of our spirit, and this is why He wants to help us with our painful problems; to unburden us, heal us and give us peace. So, get a grip with your tired hands, strengthen your weak knees, and mark out a straight path for your feet! When we call out to God our Father for help, we will not fall, even when we become weak and stumble, because God will strengthen us. So, we should not resent our trials, but rather be glad when they happen, keeping in mind as we endure them that one day we will be with Christ.

If we stop growing in Christ and stagnate in mere self-controlled morals, our growth becomes stunted and we fall into the subtle and all too common snare of damaging self-righteousness, and the pride of life.

As soon as we get our act together and start living a clean life for the Lord, the self-righteousness that is utterly against God can slink in if we begin to look down on others. This pride of life allows poisonous roots of bitterness to creep in and grow to corrupt us and those we know.

God's Spirit is glad to help us with our self-righteous bias and prejudices, whether they are personal, political or religious; we will not be disappointed as God unburdens us further and gives us more peace! Avoid this drift into an evil self-righteous attitude by **reaching out to help others**, making the tenor of our lives for encouraging others to receive God's undeserved goodness. We need to work at living a holy life in peace with others, especially with those who do not know the Lord, in hopes they someday will. We should share His gospel of peace with those we live with, and be genuinely concerned for others out of the hope that is in us. Our hope is our guarantee of living eternally with God. What we do in this life is investing for all of eternity, so we should look forward to our future with God and **live godly, self-sacrificing lives** with genuine love and concern for others.

When asked, we should always be ready to gently and respectfully explain our hope and future with God. We know **God loves us** because the Holy Spirit pours in and fills our hearts with His love.

(Psalms 19; Proverbs 11:2, 24:16; Matthew 11:30; Romans 5:1-5; Ephesians; Philippians 4:6; Hebrews 12; James 2:1-9; 1 Peter 3:15; 2 Peter 1:5-9)

Set to music as: "Abundant Life", "Abundant Life Rap", and "Abundant Life in Hawaii". Available on most streaming services.

To Fill my Spirit with God's Love

Desire and pray for moral excellence to heal my mind and body
Harmonize my spirit with God's Word and learn how to live
Have self-control and patiently keep practicing what I learn
Jesus will lift me up on my feet as many times as I need
Always be ready to explain the reason for my hope as a believer
Patiently endure trials while God is healing my spirit and character
Remember I will soon be with God, so avoid the pride of life
Help others– Get rid of my prejudices and be genuinely concerned
Let my words and inner thoughts be acceptable in your sight, Lord

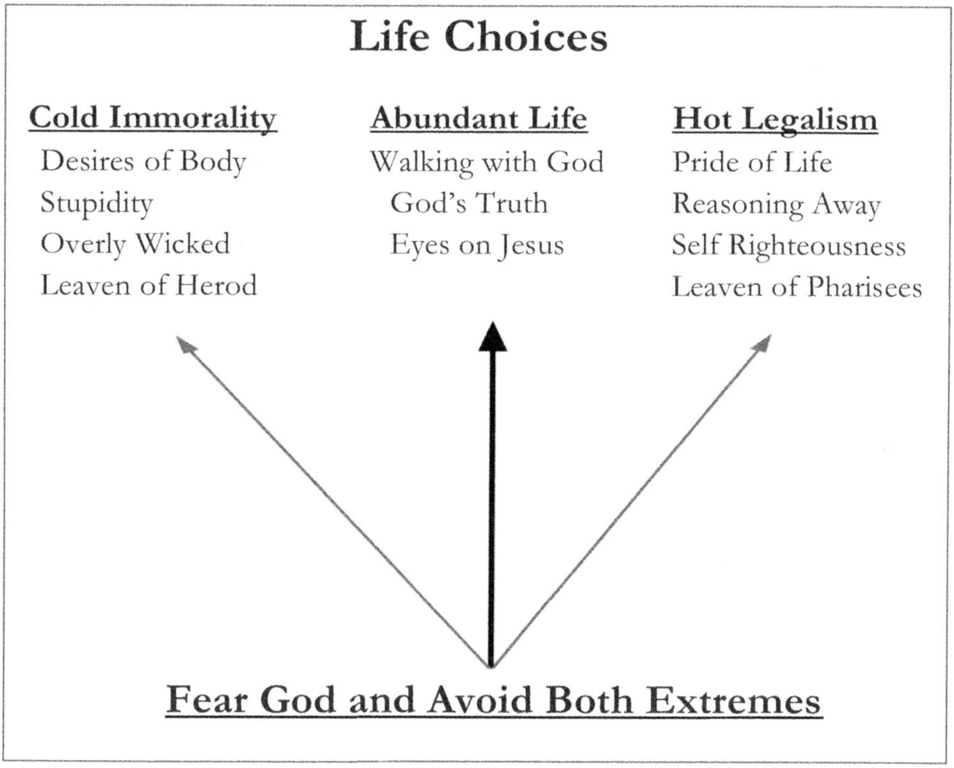

(Ecclesiastes 7:16-20; Mark 8:15; Galatians 5:16; Titus 3:8; Hebrews 12:2; 1 John 5:3)

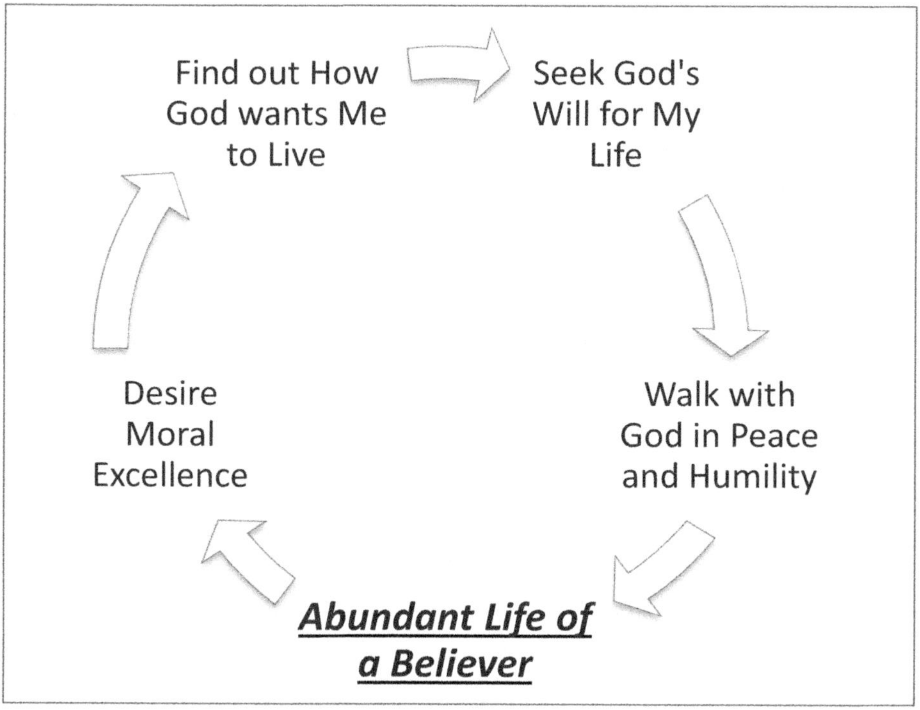

OUR HEART

God designed our heart to be our will and intellect, the place for our innermost thoughts and feelings for everything that is in us. It is in our heart that we decide to believe or turn away from God, we commune with ourselves, reason, ponder and understand. From within our heart comes our plans, integrity, joys, sorrows, and the troubles of our mind. Our heart enables us be awed, cheerful, merry or haughty, overwhelmed, humble or proud, depressed or glad, disquieted, strong or weak, desire, fail, turn back, meditate, rejoice, sorrow, praise, tremble and faint with fear, be unaware and deceived, stupid, bitter or conceited, or hardened against God and others. Our heart, whether it is weak or strong, greatly affects our imaginations and thoughts and what we do. Our heart becomes desolate when our spirit is overwhelmed. What we treasure comes from within our heart. Our spouse should be a seal on our heart. The wise in heart can have low or high IQs.

No one has a clean heart, and we lack the ability to make our own heart pure. From within our heart comes evil thoughts of murders, adulteries, immoralities, despising others, lusts, naughtiness, perversion, and hypocrisy– which gives us a double heart of deceit and mischief. We can deceitfully steal the hearts of others. Our heart is evil from our youth with its snares and nets. We can turn our heart away from the Lord, or incline our heart to the Lord. We can be slow of heart to believe, or harden our heart and not hear the Lord with an evil heart of unbelief. The fool, regardless of his IQ level, doesn't seek to understand and says in his heart there is no God, and just wants to discover the many devices within his own heart. Satan snatches God's Word away from the hearts of those who hear, but don't understand it.

God's heart grieves when our heart is evil. We do evil if we don't prepare our heart to seek the Lord. We must protect and keep our heart with all diligence, because out of it flows the issues of life. With our heart we believe to righteousness, and with our mouth confession is made for salvation. We don't have to let our heart be troubled— we can believe in Jesus, understand with our heart, and be converted with our faith. The Lord comes close to those who have a humble and broken and contrite heart. He sends His Spirit into our hearts, and we cry, "Daddy, Father!" God's Spirit dwells in our new adopted heart to comfort, teach and bless us.

We find rest for our soul when we humble ourselves and call out to the Lord and let Him know what is in our heart, and pray for Him to know our thoughts, when we set our heart to understand and purpose to walk in truth with a willing heart and mind to serve God with all our heart and soul, considering what great things He has done for us. We *can* praise and serve Him with a perfect heart and do His will.

When our heart burns within us for the Lord, and we ask with words and mediations that are acceptable in His sight, and prepare our heart to understand, He hears the desire of our humility. When we are not inclined toward our own understanding, but trust in the Lord, we allow His peace to rule in our heart. He gives us a spirit of revelation in knowledge and wisdom to walk in His statues and keep His judgments. He enlightens our understanding to know what His will and calling is.

The Lord searches our heart and already knows what is in our spirit. He perceives our thoughts and knows our hidden counsels; the secrets of our heart are known to Him. We can tell others what is in our heart, but the Lord already understands our innermost thoughts, imaginations, deceits, griefs and gladness. God's Spirit discerns the intents of our heart, and God can remove sorrow from our heart.

When we love our own heart, we commune with the Lord to seek His counsels of knowledge, wisdom and instruction so He will apply our heart to understand His will for our life. Our heart motivates us to willingly serve the Lord. We should consider what is in our heart—sometimes when our heart indicates we should do things *we* think are good, or bad, we should put these aside to obey the will of the Lord. The Lord can give us an enlarged heart of kindness and wisdom to perceive and understand His will, and the motives of others, to turn hearts towards Him. Sometimes we should settle in our heart to not mediate beforehand what we will say to others, but ask God to have His Spirit speak for us. What we fill our heart with reveals what kind of tree our fruit is from. The Lord knows the hearts of all humans, and can move our heart to understand and speak His words. The Lord calls us to love Him with all of our heart and soul and mind and strength.

Heart verse highlights are in the back of this book under references.

This HEART study is based on the HEBREW leb- will and intellect (3820); lebab- will, intellect and feelings (3824); sekviy- mind (7907); GREEK kardia- will, intellect and feelings (2588); sklerokardia- hardness of heart (4641). Strongs numbers are provided.

"Heart" is set to music on my *Strong and True* album.

Heart

Our secret thoughts and feelings
All that is in us
Flow from the gates of our heart
Our will and intellect, and beliefs and reasons
Our plans and decisions, and joys or sorrows
Our humility or pride, and cheer or depression
Our courage or fears, and desires and failures
Our bitterness or grace, and treasures or revulsions
The troubles of our mind
They all flow out of our heart.
So with all diligence protect your heart
For out of it flows our life
The decisions we make in our heart.
But we cannot make it pure.
The Lord searches and discerns
Our intents and inmost thoughts
We decide in our heart to incline to the Lord and understand.
Do not grieve God and turn to your ideologies
Those little gods you treasure in your heart
Prepare your heart and seek Him.
God comes close to a humble heart
And sends His Spirit in to dwell
To comfort and teach us
And we cry, Daddy, Father.
So with a burning heart
Ask the Lord for thoughts and words He accepts
Set your heart to understand His will for your life
And then walk in His ways
Let His peace rule in your heart
Love the Lord
With all your heart, soul, mind and strength.

"The Gates of our Heart" is a single, and "Heart" is on my *Strong and True* album

Precepts for Marriage

God made marriage a foundation of human relationships from the very beginning. Both men and women need love, respect, admiration, spending time together, doing things for each other and physical contact. Since marital conflicts arise from men and women prioritizing and valuing their needs very differently, God gave some foundational principles to help each couple understand the other's perspective so they have a great marriage.

Men Crave Respect

Men want to be the superhero swooping down in his ingeniously self-created suit of armor and saving the day, as a smile plays on the lips of his admiring, and impressed, love interest. Men also want love quite a bit, and a secure relationship, and some adventure. But mainly men require respect. A man loves a woman who is in awe of him.

Men want things to run smoothly and be able to provide for his family without fear of going in need. Deep in his spirit, he desires to be a secure provider.

Women Crave Love

Women want passionate love with her soul mate. No matter what happens in life their bonds of love will never be severed, no matter what tragedies befall them, or tempests in life beset them, she will always have her soul mate's love to count on, and she will always be loved and cherished and feel very secure in their love for each other. Women also like having fun, being admired and complimented, but they require love. A woman respects a man who loves her deeply.

Deep in their spirit women want a sheltered nest to raise her family in without fear, and a secure and close relationship with her husband.

Marriage Conflicts

All marriages have conflicts, but it is how we respond that makes the difference— Be quick to hear, overlook offenses, slow to speak, slow to anger, pursue peace, and speak the truth in a loving way.

Always respond kindly, even to criticism, and pray during conflicts.

<u>We seldom regret using kind gracious words</u>

(James 1:19; Proverbs 19:11; Ephesians 4-5; I Peter 3:9-12; Colossians 3:8-19)

Relationship Choices of the Heart

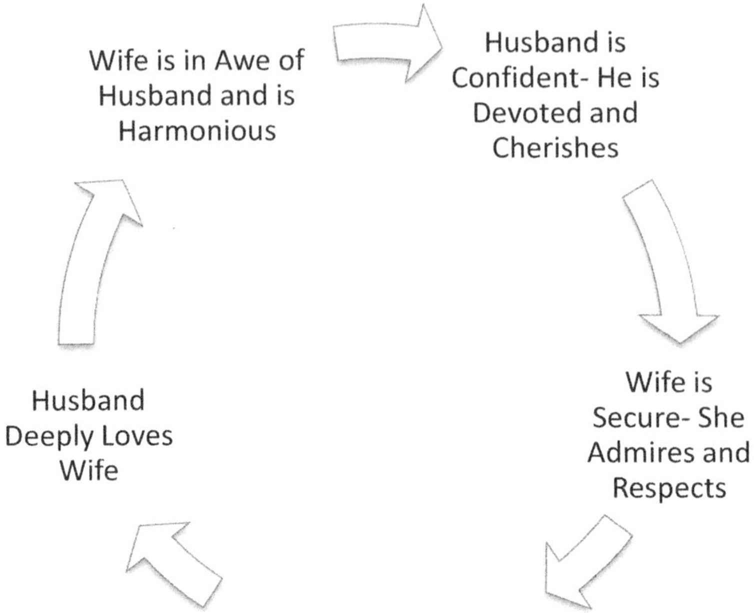

Great Marriage

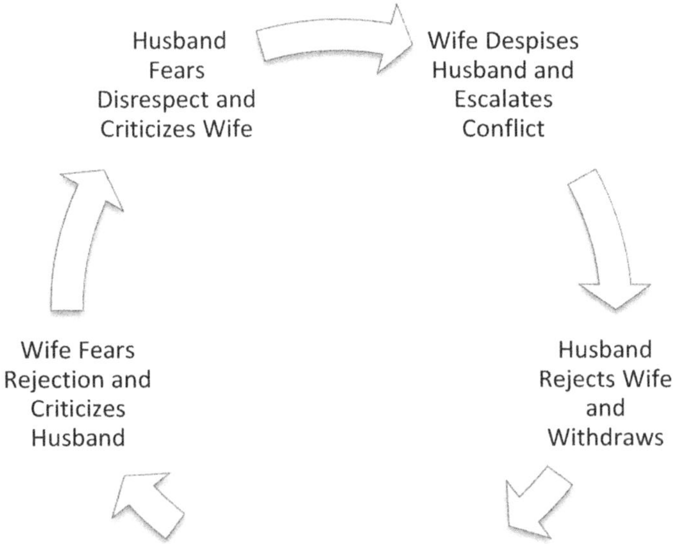

Fearful Marriage

Precepts for Raising Children

The good and bad examples set by parents are often perpetuated by their children for several generations. An amazing thing happens when someone dedicates their life and actions to Christ– the cycles of destructive generational habits can be selflessly broken in love for their own children, and they can in turn give their children a big advantage.

A big advantage for what? Do we simply want other people to admire our parenting so we feel good about ourselves? Do we just want our children to admire, love and respect us, or is it that we just don't want to worry later on about what bad things our adult children might do next?

Our motive and goal for raising kids should be to equip them to love the Lord, help others, and be productive people in our society and church, whether it be as a spouse, raising their own children, or in a career. We rightfully evaluate others based on their level of unselfishness, so shouldn't we live altruistically as an example for them? (Hebrews 12:5-11)

More than anything, children are prone to follow the example their parents set, regardless of what is said or taught.

God gives workable principles for raising kids in a harmonious, peaceful and pleasant home—

Young Children

Young children are so impressionable that their parents, especially their father, imprint on them how they will view God, so they need to know they are loved and feel secure. This will help them understand later in life that God loves them, and be secure in their relationship with God.

An angry parent is destructive and unintentionally teaches them that God is angry and vindictive and causes pain.

An absent parent teaches them God is distant and unapproachable.

A lazy parent teaches them God doesn't care what they do.

Young children need training in self-control and learn to be adept at listening to instructions. This skill set will invaluably help them later in life without their ever realizing it. It also makes for a much more peaceful home. Children who learn to behave and take care of things don't require "kid-proofed" homes, they are freed to engage their intelligence in more productive ways than with emotional conflicts. Children who exercise self-control at a young age are a joy to be with, rather than a source of stress for their parents, and for family and friends. (Ephesians 6:1-4)

Rather than teaching children to obey only after ten minutes of angry escalating conflict, teach your kids to obey with one calm request so they are pleasant to be around. Play the game of obedience—They must carefully listen so they can instantly obey what you quietly instruct them to do. Make it challenging, and compliment them for compliance. Make it fun and stop while they are still having fun so they look forward to it next time. After they learn to walk, instill work ethics into their character by showing them how to pick up trash and sweep. Heap on the compliments. They *want* to feel valued and contribute to the welfare of the home. Having them help takes longer to get a job done, but they become an invaluable help to you and others later on.

Set an example of being respectful, kind and considerate. Being unduly critical of other children can teach them to be cynical, so it is often better to just remove them from overly hostile environments.

Children need protection from exposure to harmful physical contact, substances and videos which can fester into hurtful life-changing habits that haunt them deeply as struggling adults. Children need clear boundaries, and they need to know "why" as they get older.

Children need protection from other children at times— schools and close-knit churches can create damaging hardships for kind children who are repeatedly subjected to the cruelty of other children who haven't learned how to be kind to others. (Proverbs 1:7, 3:11-12, 6:6-8,20-21, 10:4-5, 12:10-11,24, 13:4,20,24, 14:26, 20:7, 22:6, 28:13, 29:17; Colossians 4:6, James 1:20)

Older Children

Older children need to learn more of God's principles, and develop their personal relationships, work ethics and how to manage money. They especially need to know the reasons why and understand the consequences, so they can see the importance and value in what is taught.

Learning the wisdom in Proverbs will be a tremendous asset to them later in life. Read Proverbs regularly and act them out with little skits so it is fun and instructional. It will impart life-long wisdom and vital life principles. Teach them God is our "daddy." (Galatians 4:6 – Abba, Father)

God gave each child their unique spirit and soul and personality. Parents should help them discover their purpose and teach them how to pursue God's work in their life. Children are going to learn, whether it is from their parents, friends or others. If parents don't have the initiative to teach them, then they will learn from others.

Self-disciplined scholars are a by-product of developing good work ethics. Pay them for doing work, and help them manage their money by allowing them to waste it on junk. After they soon tire of their cheap toys, then casually talk with them afterwards, with the broken toys in sight, about saving money for things that last. This way, as they get older, they already know the value of money and don't waste it. Teach them about budgeting, the pitfalls of credit card debt, the benefits of good credit, new car taxes and depreciation, and pre-paying home mortgage principles.

Allot generous and regular amounts of **quantity time** to be with them so those invaluable little talks can naturally arise in conversations.

Young Adults

Emerging young adults crave to be acknowledged and respected. God wants them to become independent of their parents and dependent on Him, to care for themselves and others, to have their own family. Their desire for independence should be nurtured with respectful guidance and life lessons about the pitfalls in life, how to avoid making stupid mistakes, relationship skills, and how to be interdependent– to have others they look to for advice besides their parents. Encourage them to pursue God and relationships with others of good influence. (Ephesians 2:3, 6:4 – not to anger)

Young adults struggle if they weren't taught God's principles in relationships, work ethics and handling money when they were young.

Young adult anger and cold rejection becomes manifest in children who lacked parental love, had poor examples, or were not taught self-control. Their attitude is fueled by a rejection of their parent's failures, so disrespectfully ordering them about as if they were still young children just brings resentment. Asking them to do things and explaining the benefits of compliance will produce much better fruit in the relationship. It's never too late for parents to change and ask forgiveness for their own failures. Prayerfully repentant parents are much more respected by young adults than parents who fearfully contend to justify themselves.

Parent's Pledge– I will love the Lord with all my heart and strength.
God's word will be written in my heart so I can diligently teach the ways of the Lord to my children by telling of them as we sit in our house, as we walk and drive, at night when I tuck them in, and in the mornings.
Lord, direct what I do and let Your words be as a guiding headlamp on my forehead, and written on the walls of our home. (Deuteronomy 6:5-9)

First revere the Lord, and then learn knowledge, wisdom and instruction, which fools despise. Proverbs 1:7

The Lord's instruction, my child, do not despise, or tire of His correction, because who the Lord loves He convinces, like a father who is pleased with his child. Proverbs 3:11-12

Carry like the ant when you feel lazy, consider her ways and learn. No one tells her what to do, yet she works to provide in the summer by gathering food. Proverbs 6:6-8

My child, remember your father's instructions and do not forget your mother's guiding words. Bind them in your heart and around your neck. Where you go they lead, as you sleep they protect, and when you arise they will talk with you. Proverbs 6:20-21

Lazy hands make poverty. Diligent hands make wealth. Proverbs 10:4-5

Righteous cares for animals, but the toxic are cruel with no compassion. Proverbs 12:10

Work your land and have enough, but chasing wind lacks understanding. Proverbs 12:11

Diligent hands rule, lazy hands are forced to work. Proverbs 12:24

Lazy wants and has nothing, but the diligent have enough. Proverbs 13:4

Walk with the wise and be wise, hang out with fools and be destroyed. Proverbs 13:20

Withholding authority is hating children, but diligent instruction is kind affection. Proverbs 13:24

Those who make fun of others can't find wisdom, but those who understand learn easily. Proverbs 14:26

Righteous walks in integrity and has happy children. Proverbs 20:7

Train children the right ways and they will always remember. Proverbs 22:6

Covering sins stunts growth. So, admit when you are wrong, do right, and have compassion. Proverbs 28:13

Instructing a child brings peace and comfort and cheers your soul. Proverbs 29:17

Speak carefully with undeserved goodness and consider how to respond. Colossians 4:6

God's righteousness is not engaged while we are angry. James 1:20

What was your father like, and what do you think God is like?

Communication Choices

Use Kind Words → Acknowledge Other's Point /Apologize → Speak in Truth with Love /Forgive → **Harmonious Relationships** → Listen to Understand → Use Kind Words

Habitually Neglect Talking → Accept Mediocrity Avoid Relationships → Eat, Work and Just Try to Survive → **"Shellfish" Relationships** → Place Little Value on Others → Habitually Neglect Talking

Psalms 19 – a poem of David set to music

The cosmos speaks of splendors bright
God signing with His hand heaven's arc stretched above
Day by day pours out speech
Night by night knowledge shares
No words, no language, yet is heard
A voice streaming music
Descending to Earth
Says the sun's brilliance is set in cover
Like a groom at a wedding
Like an athlete in race rejoicing
Rising from Earth's rim revolving
Nothing is hidden from the heat
The law of the Lord is perfect, converting the soul
The testimony of the Lord is sure, making wise the simple
The statutes of the Lord are true, rejoicing the heart
The command of the Lord is pure, enlightening the eyes
The fear of the Lord is clean, enduring forever
The judgments of the Lord are true and righteous altogether
More desired are they than gold
Yeah, than fine gold
Sweeter than honey and the honeycomb
So, your servant is warned and keeping them is great reward
Who can understand his errors
Cleanse me Thou from secret faults
Keep back from me presumptuous sins
Let not them dominate me
Then I be upright and innocent of full rebellion
Let the words of my mouth and the meditations of my heart
Please you, oh Lord,
My Strength
My Refuge
My Rock
My Redeemer

"Psalm 19" is on my *Matthew* and *Psalms Rocks* albums

4

The Spirit and Super Body

Akiane Kramarik said of her "Supreme Sanctuary" painting, "In heaven I believe there is a place where each of us will be able to choose and create a sanctuary. I believe that certain things will be somewhat similar to the earth and many things completely different. This is the garden and a mansion I would like to plant and build."[14] I suspect, like me, she envisions heaven as living in our new physical bodies on a newly renovated earth with fantastic streams and waterfalls, fruit trees, and with astonishing colors in ultraviolet and infrared spectrums.

Our Earth, without a sea, has enough room for 20 acres per sanctuary for each one of an estimated 6 billion believers. Since our new physical bodies will be like Jesus had after rising from the dead, we may also fly, go through walls, teleport, and telepathically communicate. Jesus will teach us amazing things, and there will be delicious fruit to eat. No roads, cars, gas stations or dentists, just our sanctuary homes and the Holy City. There will be perfect gender equality between males and females. (Matthew 22:30) After 4.7 trillion peaceful years we will all know each other very well, and because there is no longer any sin, this is a good thing.

God, in His infinite love, freely offers us a gift pass to eternity. If we accept this, then, Jesus said, "My Father will send you the Helper, the Holy Spirit, in My name. He will teach you everything and bring to your mind everything that I have said. I will leave you with my supernatural Peace, so do not let your heart be troubled or afraid." (John 14:23-27)

After we believe, God's Holy Spirit enters and bonds with our spirit, and we become spiritually reborn children of God. His Spirit integrates with our spirit's intents to deeply influence our heart's decisions and our soul's emotions and thoughts, and allows us to commune with God.

The thoughts of our soul and the decisions of our heart are closely entwined with the intentions of our spirit, but God's living and powerful Word can discern this difference between our thoughts and our intentions. (Proverbs 4:23; 1 Thessalonians 5:23; Hebrews 4:12)

Soul Overview

The Hebrew naphash and Greek psuche refers to our soul. Our soul is entwined with our spirit, and they depart from our body after death.

Our living soul contains our personality, thoughts and knowledge, our emotions and memories, our feelings, loves and hates, what we lust for and loathe, enables us to feel elated or sad, to deceive, desire and enjoy. Our soul also interfaces with the five senses of our body and directs our actions. Our soul communes with our spirit, heart and physical body. Our soul reacts to what our spirit seeks.

We pour out our soul's emotions from the depths of our spirit— We complain of our soul's bitterness from our spirit's anguish, our soul desires what our spirit seeks, our soul is pleased with what our spirit judges to be good. (1 Samuel 1:15; Job 7:11; Isaiah 26:9; 42:1; Matthew 12:18)

Spirit Overview

The Hebrew ruwach and Greek pneuma refers to our spirit, other spirits, evil spirits, angels, or to God's Spirit. It is our spirit that interacts with the supernatural, whether it be God's Spirit or unclean evil spirits.

Our spirit is our inmost part that motivates us at deep levels to inspire our soul's feelings and thoughts. Our spirit drives our loves and hates, and influences our conscience.

Our spirit's intents influence our heart's faithfulness or instability, impels our desire for excellence or mediocrity, to be content or sojourn.

Our soul and mind contain knowledge, but our heart and spirit embody wisdom, insight and understanding. God is Spirit, so we must worship God with our spirit in truth, not in soulish emotional worship. (1 Samuel 1:15; Job 7:11; Isaiah 26:9, 42:1; Matthew 12:18; John 3:5-7, 4:23-24, 14:23)

The spirit of each believer is uniquely reborn when it spiritually bonds with God's Spirit, who upgrades our understanding and connects us to God's 3.0 spiritual network. (Genesis 3:22; 1 Corinthians 6:16, 13:13)

Analyzing the hundreds of Bible passages regarding naphash, psuche, ruwach, pneuma, leb, lebab, and kardia, and what Jesus revealed of them, yielded the Human Spirit, Heart, Soul and Body Diagram ⇨

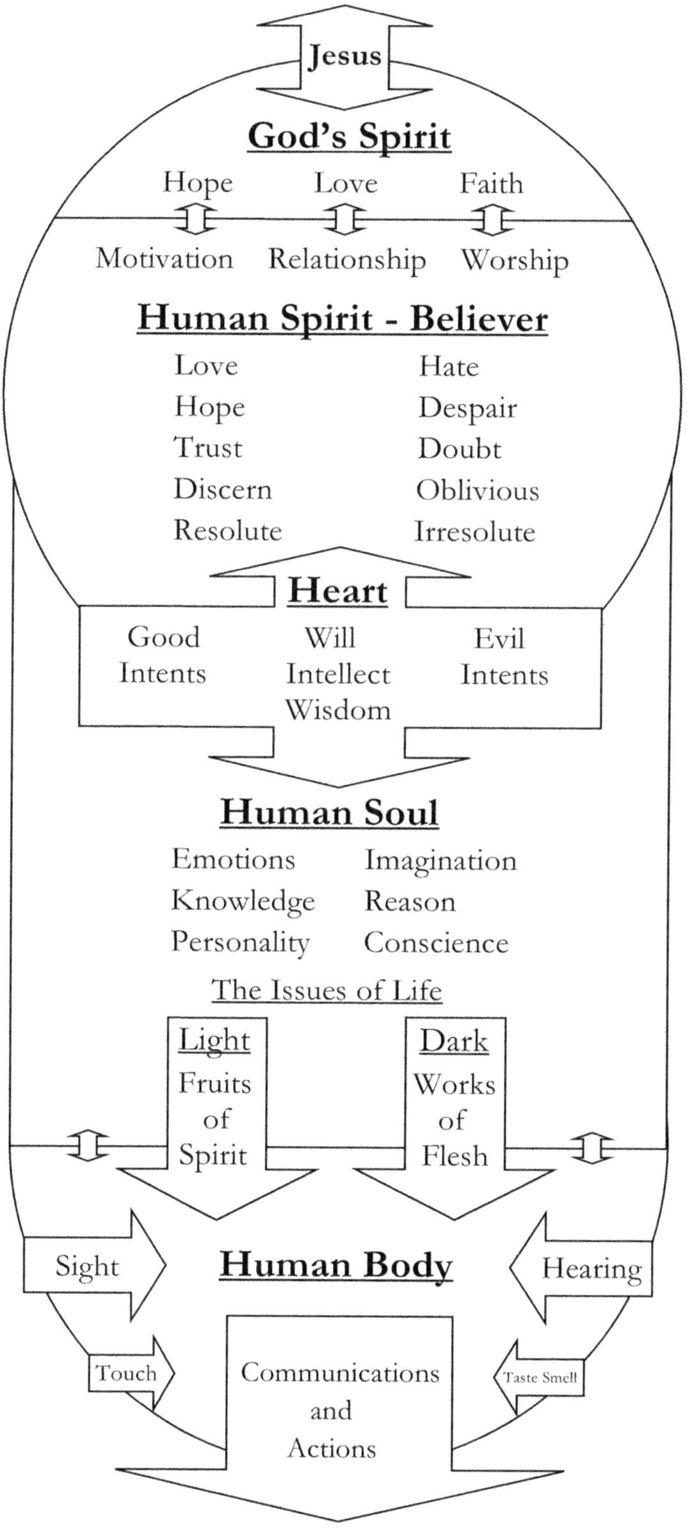
Jesus
God's Spirit
Hope
Love
Faith
Motivation
Relationship
Worship
Human Spirit - Believer
Love
Hope
Trust
Discern
Resolute
Hate
Despair
Doubt
Oblivious
Irresolute
Heart
Good Intents
Will
Intellect
Wisdom
Evil Intents
Human Soul
Emotions
Knowledge
Personality
Imagination
Reason
Conscience
The Issues of Life
Light
Fruits of Spirit
Dark
Works of Flesh
Sight
Human Body
Hearing
Touch
Communications and Actions
Taste Smell

The benefits of seeking God's will and walking in harmony with His Spirit are receiving and walking in His love with satisfaction, peace, patience, kindness, goodness, consistency, moderation and self-control. There is certainly no law against these! (Proverbs 4:23; Galatians 5:13-26)

Jesus said, "The wind blows where it wants to. You can hear the sound of it, but you cannot tell where it comes from and where it goes. So it is with everyone who is born of the Spirit." (John 3:8)

"When we tell you these things, we do not use words that come from human wisdom. Instead, we speak words given to us by the Spirit, using the Spirit's words to explain spiritual truths. But people who aren't spiritual can't receive these truths from God's Spirit. It all sounds foolish to them and they can't understand it, for only those who are spiritual can understand what the Spirit means."

Those who are spiritual can evaluate all things, but they themselves cannot be evaluated by others. For, 'Who can know the Lord's thoughts? Who knows enough to teach him?' But we understand these things, for we have the mind of Christ." - 1 Corinthians 2:13-3:1 NLV

Unbelievers can't hear God's Spirit of Truth because they are tuned in to Satan's spirit of deception, so they speak from Satan's viewpoints. God's living Spirit in believers is greater than this deceptive spirit— we have the mind of Christ and can understand this deception. We can tell if someone is listening to this spirit of deception by seeing if they can understand God's Spirit of Truth. (1 Corinthians 2:16; 1 John 4:4-6)

There is an empty place in our human spirit for God's Spirit to enter and bond and dwell within us so we become one with God. Since an unbeliever doesn't have God's Spirit, this place remains empty unless an unclean spirit resides there. An unclean spirit may help in divining stock market prices, assist an unbeliever to charismatically run a large church or corporation, or depress them. (Acts 16:16; 1 Corinthians 11:14-15)

If an unclean spirit leaves and returns to find an unbeliever's spirit is swept clean with an orderly life, it can get seven even more wicked spirits to dwell there. This may drive the person into crazed and violent acts that lead to prison or a psychiatric facility. This is how it is in our wicked generation. (Matthew 12:43-45; Acts 19:16)

Believers have a rebirth of their ***spirit*** which impacts their ***soul***.

The ***spirits*** of ***unbelievers*** are in one of these two circumstances ⇨

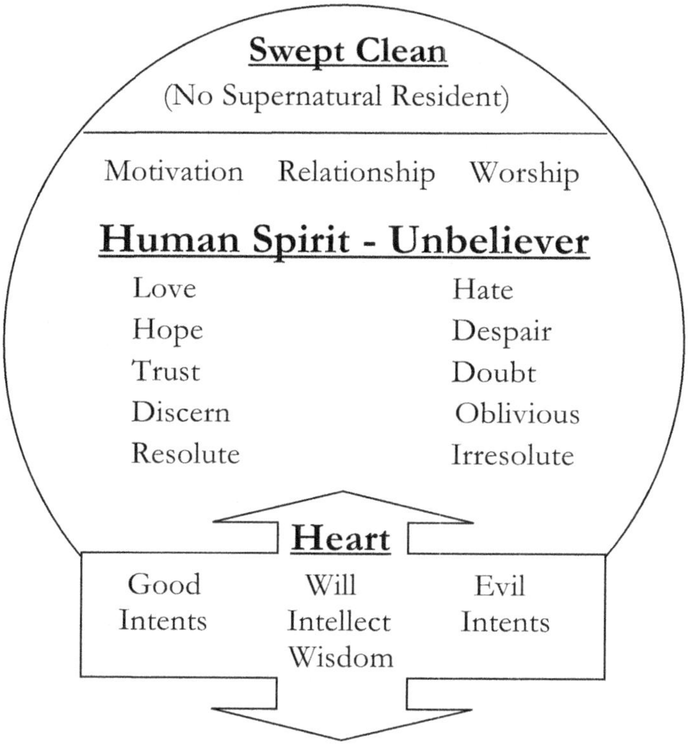
Swept Clean
(No Supernatural Resident)
Motivation
Relationship
Worship
Human Spirit - Unbeliever
Love
Hope
Trust
Discern
Resolute
Hate
Despair
Doubt
Oblivious
Irresolute
Heart
Good Intents
Will
Intellect
Wisdom
Evil Intents

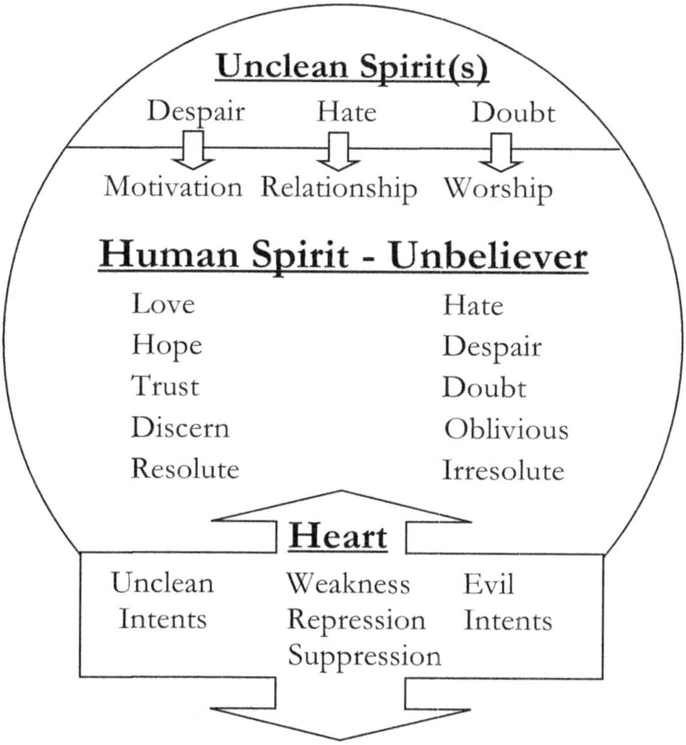
Unclean Spirit(s)
Despair
Hate
Doubt
Motivation
Relationship
Worship
Human Spirit - Unbeliever
Love
Hope
Trust
Discern
Resolute
Hate
Despair
Doubt
Oblivious
Irresolute
Heart
Unclean Intents
Weakness
Repression
Suppression
Evil Intents

SOUL

Our soul is our life that dwells within our physical body. The Lord breathed life into Adam and he became God's son and a living soul. The Lord makes all human souls, and living animal souls, and in His hand is the soul of every living thing and all human lives. When our physical body dies, our human soul and spirit goes to be with the Lord, or to Hades.

Our body exists as a place for our soul to reside, our soul is like our heart's life blood. Our soul has conscious thoughts, emotions, desires, impulses, pleasures or displeasure, sorrow, depression, grief, hate, love, and adversity. It can be enticed, angry, or satisfied. Knowledge of wisdom is a sweet reward to our soul. Our soul becomes humbled with fasting. Our physical body lusts and wars against our soul, so our soul can become unstable. Our heart interacts with and directs our soul.

Doing right preserves our soul, such as having mercy, but cruelty troubles our soul. Our soul can convince and win over the thoughts of others. We honor other people with our soul in what we do. Our soul can be physically drawn to others, and lust for others. Controlling what we say keeps our soul's thoughts from trouble— being friends with an angry man and learning his ways sets a snare for our soul's thoughts.

Jesus gave His life, His Soul, for those who believe. Our lives, our souls, are saved by believing God's word, but the Lord's Soul has no pleasure in those who draw back. When we seek to find our life we lose our soul, but when we lose our life for Christ, we find our soul. The Lord's Soul is well pleased when we love Him with all our heart, soul, mind and strength, and walk before Him in truth with all our heart and soul. Our soul, our life, rests with God. We can magnify the Lord with our soul. Sometimes we should not consider the thoughts in our soul's mind because God's Spirit will help us say the right thing. Our soul lives as the Lord lives. God can heal our soul after we sin. We convert our soul by putting God's instructions into our mind.

We should desire God with our soul's thoughts and pour out our thoughts to God, and lift our soul, our life, to God, wait upon Him and look to the Lord with delight in expectation. We should bless and praise the Lord with the thoughts of our soul and with what we do.

Soul (Life) verse highlights are in the back of this book under references.

This SOUL study is based on the Hebrew naphash (5315) and Greek psuche (5590).

"Soul" is set to music on my *Strong and True* album.

SPIRIT

God made our spirit, and all life, with the Spirit of His Word. God's Spirit conceived Jesus in Mary's womb, descended on Jesus like a dove, led Jesus to a desolate place to be tempted by Satan, raised Jesus from the dead and brought Him back to life. God's Spirit is a gift, the Spirit of Truth that is now in the world speaking through us, directing our actions, and is totally immersed within a believer's spirit.

In God's hand are the spirits of all humanity. Our spirit perceives, rejoices, and purposes with intent in what our heart wills us to do. Our spirit can have peace, trouble or distress. The sorrow of our heart can cause our spirit to be broken and changed for the better. When our spirit is overwhelmed, our heart becomes desolate. When our heart melts, so does our spirit. We can have a double portion of spirit to double our life's work. Our spirit can sustain us in illness, but we can also have a wounded spirit. Humans can have spirits of grief or heaviness, immorality or guileless, judgment, jealousy, courage, anger, sorrow, anguish, grief, corruption, trouble, faithfulness, patience, irritation with understanding, haughtiness, pride or humility. We can have an excellent spirit of understanding. **When our spirit revives, it renews our soul to stir our heart to action.**

Even if our spirit is willing to do good, our bodily desires can overcome our spirit. Our self-control arises from our heart ruling our spirit, but the foolish follow their own spirit. What we say motivates our spirit to act— wholesome words give life, but perverseness is a breach in our spirit. Fools utter all that is in their spirit, but the wise are very careful of what they say of it. We should take heed of our spirit so we are not treacherous or deceitful to our loved ones. Others can change our spirit for good or bad. A stubborn and rebellious generation have wavering spirits with hearts that are not right with God. Our spirit can turn against God, or our spirit can be contrite from a broken heart before God. No human has power over their own spirit to retain it after death— our body dies when our spirit leaves us.

God created the Universe and doesn't want sacrifice or good works, but desires, and greatly values and delights in, a gentle, humble and contrite spirit who trembles at His word. When we believe and turn to God in repentance and ask Him to create a clean heart and renew a right spirit within us, then He casts away all of our transgressions and gives us a new heart, and puts in us a new spirit so we can learn to walk in His ways. God's Spirit falls on us when we hear God's Word and we believe by faith, not by what we do. We are then born of God's Spirit and can enter His Kingdom. God sends the Spirit of Jesus as a gift down from heaven into our hearts and adopts us, and we cry, "Daddy, Father!" God's Spirit dwells within our spirit, and through Him we know we are children of God! Our body is now the temple of the Holy Spirit which lives in us, and we are not our own. God's Spirit searches our spirit and knows our inner heart and mind, and dwells in our innermost parts. God's Spirit is life, not a spirit of fear, but an earnest spirit of power and love that is spread in our hearts, a spirit of freedom and of a sound mind. God's Spirit is written in our hearts and in our spirit.

We were bought with the death of Jesus, so we should glorify God in our body and spirit, which now belongs to God, and serve God with our new spirit. But we still have our old spirit... We should seek to please our spouse with our spirit and conduct our lives and actions consistent with God's Spirit so we have a life of peace. With our soul we desire God, and with our spirit we seek God and learn to do what is right. We can live according to God's Spirit because we dwell in Him and He in us. God's Spirit comforts and guide us into all truth and gives wisdom, and directs our actions when we consciously know it, but we can decide in our heart if we will follow it. God's Spirit in us doesn't speak with enticing words of human wisdom, but with the power of God, and gives us a change of heart and a spirit of holiness, fills us with joy, and gives us discernment and strength. When we don't know how to pray, God's Spirit communicates for us with God in ways that we cannot express. We build our faith by earnestly praying in God's Spirit, and keep praying. God's Spirit speaks to us with God's Word. We call Jesus the Lord with God's Spirit.

When we suppress and quench or tempt God's Spirit by rebelling or saying bad things, it vexes and grieves God's Spirit within us. When our spirit is overwhelmed with troubles, and our soul refuses to be comforted, we can tell God about this and commune with God in our heart and search for Him diligently with our spirit, and God will direct us. He blesses us when we are poor in spirit, and does not despise a broken and contrite heart! God's Spirit regenerates and renews our spirit, purging our conscience so we can serve the living God!

When we put aside our old spirit and renew our God-given spirit, we are filled with God's Spirit. Then the Lord pours in us His Spirit of wisdom so we can know, understand and discern His will. God's Spirit gives life and freedom, and refreshes our spirit, so we should keep walking in and doing what His Spirit leads us to do, but not immoralities or legalism, which are destructive. Seeking to satisfy our bodies is corrupt, but what we do with God's Spirit lasts forever.

God's One Spirit gives gifts of wisdom, knowledge, faith, healing, miracles, discernment and languages. The Lord gives spirits of grace, discernment, craftsmanship, art, strength, inspiration, stirs our hearts to action, teaching and counsel, strengthen us, and gives an understanding spirit. His Spirit of discernment can direct and revive us. His Spirit can be on us to preach liberty from captivity to the brokenhearted.

Reverence of the Lord makes us quick to understand.

The Spirit of God dwells in all believers and joins us in unity in One Spirit as a temple of God to the Lord. So, we should endeavor to seek peace with others. When we despise other believers, we despise God, who gave us His Spirit. Other believers can refresh our spirit. We can be with others in spirit even when we are not with them. Our spirit can be restless when other believers are troubled. We should stand with other believers and worship God with our spirits. We are all in God's Spirit, striving together for the faith of the gospel. In God's Spirit the gospel comes in power. Treat others who sin in a spirit of meekness, and develop habits of earnest prayers for ourselves and others so that God will preserve our whole spirits and souls and bodies to be blameless when our Lord Jesus Christ comes. Our new spirit groans for a new, refabricated body, but our spirit is perfected by God, so we should not be shaken in our spirit or mind by words, but allow God's Spirit to give us grace, wisdom and His knowledge.

God makes His angels spirits of flaming fire and sends them out to minister to us. God can make unclean or strange spirits pass from the land, or send a perverse or lying spirit into the midst of a country and cause them to err in everything they do. People can be vexed with unclean spirits, but God's Spirit can control these evil spirits.

Spirit verse highlights are in the back of this book under references.

This SPIRIT study is based on the Hebrew ruwach (7307) and Greek pneuma (4151).

"Spirit" is set to music on my *Strong and True* album.

Spirit DNA

A tiny planted seed growing into a productive new plant is like a believer's spirit and soul growing into a powerful new body. Jesus said, "A grain of wheat is just one grain of wheat, but if it gets into the ground and dies, it will grow and produce many grains."

A modern allegory to a single grain of wheat is DNA, but a special sentient spirit DNA that interfaces with our soul's emotions, memories, thoughts, imagination, conscience and personality.

(Job 7:11; Isaiah 42:1; Matthew 12:18, 13:30; John 12:24-26; I Corinthians 15:37-45; 1 Thessalonians 5:23; Hebrews 4:12; John 12:24)

Jesus will protect our soul if we believe in God and ask in repentance for help and forgiveness. We become spiritually reborn when God's Holy Spirit bonds with our spirit's DNA. The Holy Spirit then dwells inside each believer, comforting and teaching us. This gift can't be earned by simply living a good life and trying not to sin. Only the indwelling of God's Spirit protects our spirit DNA and guarantees that we will someday receive new physical bodies. A believer's spirit DNA returns to God after the death of our body, because it is now inseparable from God's Spirit. But the spirit of an unbeliever remains corrupted by sin after death because they lack the power to resurrect themselves, and their soul and spirit later returns to God for judgment at the end.

(Ecclesiastes 12:7; Revelation 20:4-5,11-15)

Since God is God of the living, not of the dead, God will someday match each believer's soul and spirit to an incredible new physical body.

(Matthew 13:30; Mark 9:42-44, 12:27; Romans 3:23, 8:23; 1 Corinthians 15:35-58)

Our soul is designed to dwell in and interface with a physical body, so after death our soul and spirit will desire another one. Jesus will transform us when He is revealed, and we will acquire powerful new spiritual bodies of majestic splendor. We don't really know what our new bodies will be like, but we do know that our bodies will be like His.

(Matthew 22:29-32; Luke 24:31-39; Acts 1:9; 1 Corinthians 15:37-44; I John 3:2)

But, before we get our new physical superbodies, what exactly happens to our soul and spirit after our corrupt physical bodies die? Where is Paradise, Heaven, Hades and Hell? Why did Jesus have to die on the cross, and what changed afterwards? Where did souls go before Jesus died on the cross, and where do they go to now? Who controls Hades– Satan or Jesus? Abaddon?

Overview of the Spirit Investigation

Where souls and spirits have gone, and go to now, changed with the crucifixion and physical resurrection of Jesus.

Before the Death of Jesus Christ: All souls went to a domain in the lower parts of Earth. This domain is split in two by a deep fissure, with Hades on one side and Paradise (aka Abraham's Creek), on the other. Hades is unpleasant and Paradise is pleasant. When Satan gained the power of death, he gained some control over this domain as the god of this world. Abaddon is in Hades. The souls of believers became trapped in Paradise and had to wait for Jesus to retroactively die for their sins to rescue them. God wanted to get His people out of Paradise.

Death of Jesus Christ: Jesus died and went down and emptied out Paradise and brought with Him all of these souls up to heaven. The souls in Hades were left behind to wait for their final judgment of being cast into the lake of fire, which hasn't happened yet.

After the Resurrection of Jesus Christ: Believers now skip Paradise and go directly to heaven to be with Christ. The souls of those blotted out of God's Book of Life are still being added to Hades.

End Time: Abaddon is cast into the Lake of Fire. Later, those in Hades are brought up and judged. God checks the Lamb's Book of Life, and everyone's name that has been blotted out is cast into the lake of fire and to their second death, which is either ceasing to exist, or burning forever like Satan and the false prophet. Those with Christ will be here on the newly renovated earth with no mountains or seas.

The Spirit Investigation

Before the Death of Jesus Christ

King Saul is in a time of great distress and uncertainty and God won't answer him, so King Saul breaks his own law and asks a witch to call up the spirit of Samuel to get his advice. The soul of Samuel rises from the ground. Samuel is annoyed with Saul for disturbing him—

"After Samuel died and they lamented and buried him, Saul put all the mediums and spiritists out of the land. Then the Philistines gathered and encamped at Shunem. So Saul gathered all Israel together and they encamped at Gilboa. But when Saul saw the army of the Philistines, he was afraid and his heart trembled greatly.

Saul inquired of the Lord, but the Lord did not answer him by dreams or the Urim or by the prophets. Saul said to his servants, 'Find me a woman who is a medium so I can ask her.'

His servants said, 'In fact, there is a woman who is a medium at En Dor.'

Saul disguised himself by putting on other clothes, and went with two men at night to the woman, and asked her, 'Please, conduct a séance for me, and bring up the one I will name to you.'

The woman said, 'Look, you know what Saul has done, how he has cut off the mediums and the spiritists from the land. Why are you laying a snare for my life, to get me killed?!'

Saul swore to her, 'As the Lord lives, you won't be punished.'

So the woman asked, 'Who shall I bring up for you?'

He said, 'Bring up Samuel for me.'

When the woman saw Samuel, she cried out, and said, 'Why have you deceived me? For you are Saul!'

King Saul said, 'Do not be afraid! What did you see?'

The woman said, 'I saw a spirit ascending out of the earth.'

Saul asked, 'What is his form?'

She said, 'An old man with a mantle is coming up.'

Saul perceived it was Samuel and bowed with his face to the ground.

Samuel said, 'Why have you <u>disturbed</u> me by bringing me up?'

Saul answered, 'I am deeply distressed; for the Philistines make war against me, and God has departed from me and does not answer me anymore, neither by prophets nor by dreams. So I called you so you can reveal to me what I should do.'

Samuel said: 'Why do you ask me, seeing that the Lord has left you and become your enemy? The Lord has done what I told you He would do, the Lord has torn the kingdom out of your hand and given it to your neighbor, David, because you did not obey the Lord. The Lord will deliver Israel and you into the hand of the Philistines. Tomorrow, you and your sons will be with me.''' - I Samuel 28:3-19

Samuel's soul looked like him, and could still see, hear, speak, and remember. Where did Samuel's soul rise from, and go back to? Jesus explained where this was with His story of a rich man in Hades who sees poor Lazarus in Paradise. This also offers a plausible reason for why Samuel was annoyed with Saul for disturbing him—

"There was a certain rich man who was clothed in purple and fine linen and fared sumptuously every day. But there was a certain beggar named Lazarus, full of sores, who was laid at his gate, desiring to be fed with the crumbs which fell from the rich man's table. Moreover the dogs came and licked his sores. So it was that the beggar died, and was carried by the angels to Abraham's Creek. The rich man also died and was buried. And being in torments in Hades, he lifted up his eyes and saw Abraham afar off, and Lazarus in his bosom.

Then he cried and said, 'Father Abraham, have mercy on me, and send Lazarus that he may dip the tip of his finger in water and cool my tongue; for I am tormented in this flame.'

But Abraham said, 'Son, remember that in your lifetime you received your good things, and likewise Lazarus evil things; but now he is comforted and you are tormented. And besides all this, between us and you there is a great gulf fixed, so that those who want to pass from here to you cannot, nor can those from there pass to us.'

Then he said, 'I beg you therefore, father, that you would send him to my father's house, for I have five brothers, that he may testify to them, lest they also come to this place of torment.'

Abraham said to him, 'They have Moses and the prophets; let them hear them.'

And he said, 'No, father Abraham; but if one goes to them from the dead, they will repent.'

But he said to him, 'If they do not hear Moses and the prophets, neither will they be persuaded though one rise from the dead.'"
- Luke 16:19-31

Jesus confirmed Paradise is a real place just before He died, when one of the criminals hanging next to Him impiously said, "If You are the Christ, save Yourself and us!"

The other criminal rebuked him, saying, "Do you not fear God, even after you got the death sentence!? We deserve death, but He has done nothing wrong!" Then he said to Jesus, "Lord, remember me when You come into Your kingdom."

Jesus replied, "I assure you, today you will be with Me in Paradise."

The thief had a short stay in Paradise, leaving within 48 hours to depart with Jesus.[16] (Luke 23:39-43) ("Paradise" is on my *Peace and Power* album)

After Bathsheba's affair with King David, she delivered a baby boy, who became very sick. David wept and fasted in hopes that the Lord would relent and make the baby well. But after the baby died, he said, "I fasted and wept when the child was alive, because I thought, 'Who can tell if the Lord will be gracious to me so the child will live?' But now that he is dead, so why should I fast? Can I bring him back again? I will go to him, but he will not return to me." (2 Samuel 12:23)

Where did David and the baby go after they died? The only place mentioned is Paradise (Bathsheba later had another son, Solomon, who was incredibly wise and wrote Song of Songs, Ecclesiastes, and most of Proverbs, so God brought good from this ugly situation).

Death of Jesus Christ

Christ went down and emptied Paradise, or Abraham's Creek, of its captive souls, and they were finally able to go with Him to heaven—

Jesus said, "I led captivity captive, and gave gifts to men." In other words, after Jesus died on the cross, He went into the heart of the earth and rescued all of the believers who were being held captive in Paradise, those who had died after Satan gained control of death, and Jesus led these captives out and up to heaven. Now, Jesus freely gives His gift of His Spirit to any who believe and accept this undeserved goodness—

"But to each one of us grace was given according to the measure of Christ's gift. Therefore He says: "When He ascended on high, He led captivity captive, and gave gifts to men." (Now this, "He ascended" - what does it mean but that He also first descended into the lower parts of the earth? He who descended is also the One who ascended far above all the heavens, that He might fill all things.)" - Ephesians 4:7-10 (Psalms 68:18-20; John 4:10; Acts 2:38, 10:45; Romans 6:23; 1 Corinthians 7:7, 13:2; Ephesians 2:8; James 1:17; 1 Peter 4:10)("Paradise" is on my *Matthew* album)

"For as Jonah was three days and three nights in the belly of the great fish, so will the Son of Man be three days and three nights in the heart of the earth. The men of Nineveh will rise up in the judgment with this generation and condemn it, because they repented at the preaching of Jonah; and indeed a greater than Jonah is here. The queen of the South will rise up in the judgment with this generation and condemn it, for she came from the ends of the earth to hear the wisdom of Solomon; and indeed a greater than Solomon is here." – Matthew 12:40-42

Abraham, Samuel, David and the souls of all other believers who died before Jesus died on the cross were being held captive in Abraham's Creek, or Paradise, waiting until Jesus retroactively died on the cross for their sins so He could then come down to rescue them and take their souls and spirits back with Him and up to Heaven (Hebrews 9:26-28)—

"For God presented Jesus as the sacrifice for sin. People are made right with God when they believe that Jesus sacrificed his life, shedding his blood. This sacrifice shows that God was being fair when he held back and did not punish those who sinned in times past, for he was looking ahead and including them in what he would do in this present time. God did this to demonstrate his righteousness, for he himself is fair and just, and he declares sinners to be right in his sight when they believe in Jesus." - Romans 3:23-26 NLT

After the Resurrection of Jesus Christ

Jesus now holds the keys of Death and Hades—

After Jesus was resurrected, He put his right hand on John's shoulder, and said, "Do not be afraid; I AM the First and the Last. I AM He who lives, and was dead, and I AM alive forevermore. Amen. And I have the keys of Hades and of Death." - Revelation 1:17-19.

Since Jesus now holds these keys, the gates of Hades can no longer prevail against believers. But those whose names were blotted from God's Book of Life are still in Hades waiting for their final judgment in the lake of fire. (Matthew 16:18; Revelation 20)

Jesus emptied Paradise when He took believers to Heaven with Him. The souls of believers who currently die now skip Paradise and go directly to heaven to be with the Lord. Paul writes, "So we are always confident, knowing that while we are at home in the body we are absent from the Lord. For we walk by faith, not by sight. We are confident, yes, well pleased rather to be absent from the body and to be present with the Lord." - 2 Corinthians 5:6-8

Paul explains that he is going straight to heaven when he dies: "For to me, to live is Christ, and to die is gain. But if I live on in the flesh, this will mean fruit from my labor; yet what I shall choose I cannot tell. For I am hard-pressed between the two, having a desire to depart and be with Christ, which is far better. Nevertheless to remain in the flesh is more needful for you." - Philippians 1:21-25

Pop Quiz on Hell

Some confusion over what Hell exactly is might be due to some Bibles translating the Hebrew "sheol" as "Hell." But "sheol" can mean pit, grave, death, Hades, Paradise, or may refer to both Hades *and* Paradise as the place of the dead. Sheol, in Jonah's case, was inside a big fish. Some translations simply leave it as "sheol" to avoid bias, which allows us to sort out what each "sheol" means by the context of the passage.

Hades and the lake of fire are real places, but there is actually no third place specifically called "Hell." See if you can determine if it is Hades or the lake of fire that is being referenced in these verses—

Question 1: Jesus said, "But whoever causes one of these little ones who believe in Me to stumble, it would be better for him if a millstone were hung around his neck, and he were thrown into the sea. If your hand causes you to sin, cut it off. It is better for you to enter into life maimed, rather than having two hands, to go to hell, into the fire that shall never be quenched - where 'Their worm does not die, And the fire is not quenched.'" (Mark 9:42-44)

Answer: Jesus is quoting from Isaiah 66:24, which refers to the Lord's future millennial reign that occurs before the last judgment with the lake of fire in Revelation, so Jesus is referring to Hades. Their soul becomes a corrupted sin "worm" in Hades, but does not die.

Question 2: Isaiah wrote of the king of Babylon in Hell—

"Hell from beneath is excited about you, to meet you at your coming; it stirs up the dead for you, all the chief ones of the earth; it has raised up from their thrones all the kings of the nations.

They all shall speak and say to you: 'Have you also become as weak as we? Have you become like us?'

Your pomp is brought down to Sheol, and the sound of your stringed instruments; the maggot is spread under you, and worms cover you." - Isaiah 14:9-11

Answer: These kings are in Hades, also referred to here as Sheol. Souls can see, hear and speak in Hades.

Question 3: Jesus said, "Then the King will turn to those on the left and say, 'Away with you, you cursed ones, into the eternal fire prepared for the devil and his demons. For I was hungry, and you didn't feed me. I was thirsty, and you didn't give me a drink. I was a stranger, and you didn't invite me into your home. I was naked, and you didn't give me clothing. I was sick and in prison, and you didn't visit me.'

Then they will reply, 'Lord, when did we ever see you hungry or thirsty or a stranger or naked or sick or in prison, and not help you?'

And he will answer, 'I tell you the truth, when you refused to help the least of these my brothers and sisters, you were refusing to help me. And they will go away into eternal punishment, but the righteous will go into eternal life.' (Matthew 25:41-46 NLT)

Answer: Jesus describes a future everlasting fire prepared for the devil and his angels, which Revelation 20:15 explains is "the lake of fire," where the souls that are now in Hades will also be thrown in to, which will be their second death. Jesus said, "Do not fear those who kill the body but cannot kill the soul. But rather fear Him who is able to destroy both soul and body in hell" (Matthew 10:28; Revelation 21:8)

Question 4: David said, "I will bless the Lord who has given me counsel; My heart also instructs me in the night seasons. I have set the Lord always before me; Because He is at my right hand I shall not be moved. Therefore my heart is glad, and my glory rejoices; My flesh also will rest in hope. For You will not leave my soul in Sheol, nor will You allow Your Holy One to see corruption." (Psalms 16:7-10)

Answer: Sheol is the place of the dead divided by a deep chasm into Hades and Paradise. David is glad that the physical body of the Lord Jesus will not decompose, and the Lord will not leave David's soul in the Paradise part of Sheol. Acts 2:30-34 and Psalms 49:15 clarifies that Jesus was resurrected with a new spiritual body, and that David was referring to his soul's redemption from Sheol *after* the death of Christ.

Question 5: Christ proclaimed the truth "to the spirits in prison, who formerly were disobedient, when once the Divine longsuffering waited in the days of Noah, while the ark was being prepared." (1 Peter 3:19-20)

Answer: It is Tartarus, the bottomless pit, the Abyss of Hades, the place of darkness where the punished angels are kept who had human offspring in the days of Noah. The inhuman destroying angel Abaddon will be released from Tartarus during the Great Tribulation. Abaddon is also known as Apollyon, which associates him with the Apollon of renowned Greek mythology. Tartarus is also in Greek mythology.
(Genesis 6:4; Proverbs 27:20; Luke 8:31; 2 Thessalonians 2:8; 2 Peter 2:4; Jude 6; Revelation 9:11 DRB, 11:7, 17:8)

Enoch, Moses and Elijah

Enoch, Moses and Elijah are three exceptions to where souls have gone to in the past, and they seem to play a part in Earth's future—

God buried Moses after he died. Satan disputed the body of Moses since he controlled death at the time, and souls were supposed to go to either Paradise or Hades after their bodies died. Michael the archangel, while contending with Satan over the body of Moses, didn't make any reviling accusation, but simply said, "The Lord rebuke you!"

Elijah and Enoch went directly to God without physically dying. Their souls and spirits went to heaven, but the circumstantial evidence also involves their physical bodies—
(Genesis 5:24; Deuteronomy 34:6; 2 Kings 2:11; Jude 9)

Hundreds of years after they left Earth, both Elijah and Moses physically surface six days before Jesus died on the cross. Jesus led Peter, James and John up on a high mountain and they saw Jesus transfigured. His face became as bright as the sun and His clothes radiated with light. Moses and Elijah then appeared with Jesus and they discussed something.

Moses and Elijah might be God's two witnesses during the Great Tribulation. An energy force emanates from their mouths to consume any who try to harm them. They also prevent rainfall, turn the rivers and oceans into blood, and strike plagues on Earth whenever they want. Then Abaddon, the powerful angel beast who emerges from the bottomless Abyss, overcomes and kill them. But no one is allowed to bury them. All the people of Earth gloat and celebrate their deaths by merrily exchanging gifts. But after three and a half days God's Spirit of life enters them, they stand on their feet, and a great voice says, "Come up here!" All the humans of Earth watch aghast in horrified terror as they rise and disappear into a cloud. (Matthew 17:1-4; Revelation 11:3-12)

The souls of believers now die skip Paradise and go directly to be with the Lord, as Moses evidently did. Believers of the future will physically go to the Lord without dying, as Elijah and Enoch did.

Enoch, who God also translated alive into heaven, said, "The Lord is coming with a vast multitude of holy ones to judge the people of Earth and convict them of all the ungodly things they have done, and for all of the ungodly insults they said about God."
(Malachi 4:5-6; Mark 9:2-13; Luke 1:17; John 1:19-23; 1 Corinthians 15:52; 2 Corinthians 5:6-8; 1 Thessalonians 4:16-17; Hebrews 11:5; 1 John 3:2; Jude 14-15)

The Soul Flowchart

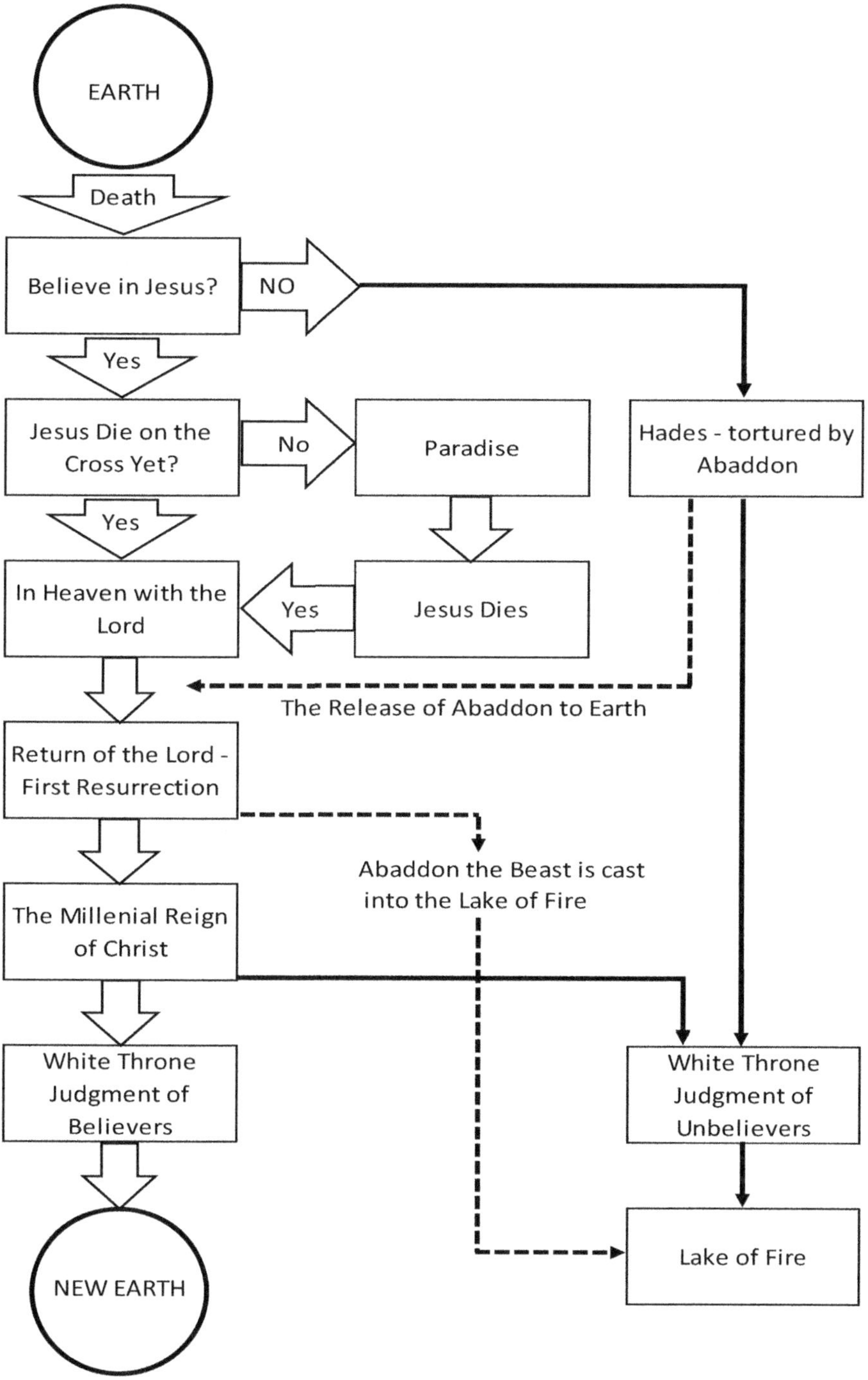

The Eternal Flame

The seven eternal Spirits of God burn before Him in Heaven. God told the Israelites to continually burn a consuming flame as a copy of what is in Heaven, and to use this flame when they offered sacrifices for their sins. The Temple Mount, on a hill overlooking Jerusalem, became the place where this continual flame, or daily flame, was always kept burning for sacrifices.

(Exodus 27:20, 29:42; Leviticus 6:8-13; Hebrews 8:5, 12:29; Revelation 4:5, 15:5)

This continual flame was temporarily taken away two times. The first was in 586 BC when Jerusalem fell and the temple was torn down. Soon afterwards, Ezekiel detailed the new temple design the nation of Israel was to use to rebuild it with if they were ashamed of all they had done; it also was to have blood sacrifices offered for their sins. Haggai 1 tells of the Lord's anticipation when the time came to build this grand new temple. But, the Jews were not ashamed, had intermarried with pagan women, practiced usury on fellow Jews, and even some priests were defiled, so Ezekiel's grand temple was never built. When Sheshbazzar laid the new temple foundation in 516 BC, it was even smaller than the first one, and the older Jews wept. (Ezra 3:12, 5:13-16, 9:1-15; Nehemiah 2:10, 5:3-13, 6:14, 13:4-30; Ezekiel 40:4, 43:10-11; Haggai 1, 2:3)

The second time the continual flame was taken away was prophesied of in Daniel 8:10-14 and 11:33-32, when Antiochus Epiphane cast truth to the ground and set up an altar to Zeus in the place of the sanctuary in 167 BC. The continual flame was cast down and the temple was trodden on for 2300 mornings (Hebrew "boqer") and nights (Hebrew "ereb"). But 1,150 solar days later, in 165 BC, courageous Jews won their temple back with daring exploits and cleansed it in a rededication. The Jewish Hanukkah festival observes this rededication, when only one day's supply of oil miraculously burned for eight. (1 and 2 Maccabees)

Only the Western Wailing Wall is left standing after the temple was torn down a third time in AD 70. The Temple Mount site was in disuse until the Dome of the Rock shrine was completed in AD 691— Daniel's 11:36-45 prophesy may be referring to this weighty Muslim dominion.

The taking away of the continual (daily) flame of sacrifice isn't mentioned at all in Daniel's 9:24-27 prophesy, but rather that Jesus is to be anointed as the King at the end of the seventieth week, but before this, halfway through that seventieth week, **the blood sacrifice and**

bloodless sin offerings would cease. This was fulfilled by Jesus, who made all blood and daily sacrificing **obsolete** by sacrificing His own life for our sins. Since Jesus permanently ended all blood *and* daily sacrifices, how is God's continual flame represented here on Earth now? (Hebrews 7:23-28, 8:1-13, 9:1-28, 10:18) When the Holy Spirit came on Pentecost, cloven flame-like lights appeared above their heads. Jesus says we are the light of the world, and a light set on a hill cannot be hid— Believers are the temple, light, living sacrifice, priests and kings who now bear witness of God. (Matthew 5:14-16; John 3:19, 5:35-36; Acts 1:8, 2:1-8, 32; Romans 12:1; 1 Corinthians 3:16; Ephesians 5:8; Revelation 1:6, 1:20)

Daniel 12:11-12 prophesies the continual (daily) flame will be taken away for 1,290 "yohms", and blessed are those who wait for 1,335 "yohms". "Yohms" can mean a time of "days" or "years". This prophesy is now unclear, but it may correlate to God's two lampstands burning His flame before Him in heaven, who become God's two witnesses on Earth for the 1,260 days during the tribulation as the saints are overcome by the beast angel Abaddon, who also later overcomes these two witnesses. (Zechariah 4:11-14; Revelation 11:3-5, 13:7-8, 18:24)

In the very end, there will no longer be any temple at all in God's new city, because God the Father and Jesus will dwell with us and be the temple, and Jesus will be the light. (Revelation 21:22-23, 22:5)

The Antichrist

The Greek "antichristos" is mentioned *only* in 1 John 2:18,22, 4:3, and 2 John 7. John explains the antichrist that they heard was coming and is already here are those who deny the Father and the Son and don't confess that Jesus came in the flesh. The angel beast and the false prophet of the tribulation *are* antichrist, *but* the Greek "antichristos" is never actually mentioned or used at all in Revelation! (Daniel 7:11; Revelation 19:20)

The passages about Antiochus Epiphane (Daniel 8:23-26 and 11:30-35) likely serves as an analogy of the false prophet of Revelation 13:11-18, but Daniel 8 and 11 prophesied of Alexander the Great and how his kingdom was to be broken into four smaller kingdoms, and of the events after Alexander— These later events depict many different nations and kings encompassing hundreds of years, and are symbolized as the king of the North and the king of the South. (According to legend, Jewish leaders deterred an invasion by greeting Alexander and showing him the passages in Daniel that prophesied of his greatness and coming to them.)

Heaven's Angels

Do angels dwell on Earth, or in Heaven, or somewhere in between? Where exactly *is* Heaven? There are three different heavens:

1) The first heaven is our atmosphere. (Acts 14:17; Deuteronomy 28:12)

2) The second heaven is the work of God's fingers, the God ordained Universe of planets, stars, constellations and galaxies. Earth is God's footstool and the Universe is His throne, His house. God has many dwelling places among the stars, but God the Father doesn't dwell on Earth or in the far reaches of the Universe. (Deuteronomy 4:19; 1 Kings 8:27; Psalms 8:3; Isaiah 13:10, 66:1; John 14:2; Acts 7:49; I Corinthians 15:40)

3) God the Father dwells in a third heaven beyond the Universe in timeless eternity. The Ancient of Days, the high and lofty Holy One, said, "I live in the high and holy place but also with those whose spirits are contrite and humble, and where there is an assembly in the far reaches of the north." God's Son interacted directly with humanity and Satan throughout the Old Testament. Born as Jesus, he was made a little lower than the angels, killed and resurrected, and ascended to the third heaven for us, and sat down at the right hand of God the Father, at the throne of the Majesty. Stephen, just before his martyred death, saw Jesus standing in the third heaven next to God the Father.
(Job 1:6-12; Daniel 7:9,13,22; Isaiah 14:13,57:15; Acts 7:55; 2 Corinthians 12:2; Ephesians 4:10; Hebrews 2:7, 8:1, 9:24)

The angels of children beholding the face of God the Father are in this third heaven, where a hundred million angels surround God's throne.
(Psalms 113:4-6; Daniel 7:10; Matthew 18:10; Colossians 2:18; Revelation 7:11)

Some of heaven's angels come to Earth. Angels are spirits, and can also have visible, physical and powerful superhuman embodiments. They can blind humans, cause mass deaths, and shine like lightning. It was an angel who stirred the Pool of Bethesda and healed the first human to step into the water. Some believers have unknowingly invited strangers, who were actually angels, to eat and stay in their home. Jesus will send His angels to gather His elect from one end of heaven to the other. Jesus will come back to Earth with His angels. Angels have their own languages. Believers can be a good or bad example for the angels. The Devil has his own angels as well. Humans are a little below the angels.
(Genesis 19:11, 28:12; 2 Kings 6:17, 19:35; 1 Chronicles 21:15; Psalms 104:4; Daniel 9; Matthew 16:27, 24:31, 25:41, 28:3; Luke 1:19; John 1:51, 5:1-4; I Corinthians 11:10, 31:1; Hebrews 1:13-14, 2:7-9, 13:2)

Stars – Celestial Bodied Beings

Stars are symbols of power, like a scepter. Jesus is the bright morning star. Angels are also stars, like the seven stars, or angels, of the seven churches, or the stars from heaven who fought for Israel. God walks on the vault of the lofty stars of the Universe and established ordinances among the stars, and **set their dominion over Earth**, but God can seal up these stars. Some stars do not dwell in eternity with God. (Genesis 37:9, Exodus 32:13; Numbers 24:17; Deuteronomy 1:10; Judges 5:20; Job 9:7, 22:12-14, 38:32,7; I Corinthians 15:38-40; Revelation 1:16-20, 2:28, 8:13, 12:1, 22:16)

The Lord can cast down those who reside in the stars— Lucifer, the day star and son of the dawn, fell to Earth. Also, some angels did not keep their first estate but left their own habitation and shamefully wandered to Earth and sinned. These wandering stars are now locked in utter darkness. Israel may have worshipped some of these wandering angels, such as Moloch and Rephan.
(Genesis 6:4; Isaiah 14:12-13; Matthew 25:31; Acts 7:42-43; Jude 1:6-13)

Will the dangerous wayward celestial bodied beings having open human contact during the Great Tribulation be worshiped as evolutionarily advanced extraterrestrials? If so, will some falter in their faith?— When Michael and his angels cast Satan, the great red dragon who controls seven heavenly crowned heads, and his angels down to Earth, it is one third of the stars, or angels, who are cast down with Satan. Since no place is found for them, they fall spectacularly to Earth like a fig tree casting its unripe figs when it is shaken by a strong wind.

Those who dwell in the stars rejoice when Satan and his angels are cast to Earth, but those living on Earth are warned of the woes to come— Wormwood, a great star, comes blazing through our atmosphere like a radioactive torch-lit spaceship and toxically pollutes a third of Earth's water, causing many deaths. A great shadow blocks a third of Earth's sky. Then, a fallen star is given the key to the shaft of the abyss, which could be deep within the Earth or in space, but it has a shaft leading to Earth. The portal to this abyss is opened and Abaddon, the angel king of the Abyss, a dangerous and highly intelligent beast, emerges from the dense, black smoke with his army of weaponized creatures, and they torment humanity. Then, four angels and their formidable flying army of two-hundred million flying extraterrestrial crafts are released.
(Isaiah 34:4; Daniel 8:10; Amos 5:8; Obadiah 4; Mark 13:25; Revelation 1:18, 6:13, 8:10-13, 9:1-19, 12:3-13, 13:1, 17:8)

ATDs - Angelic Transportation Devices

In the 60 Minutes broadcast, "Navy pilots describe encounters with UFOs", Luis Elizondo and Christopher Mellon explain how they quit their high-level government jobs to help get UFO evidence before Congress and to the general public. Evidence such as jet pilot lieutenant Ryan Graves saying UFOs were observed every day by jet pilots for more than two years off the coast of Virginia. These disclosures led to the Pentagon releasing a report on UFOs in June 2021. Humanity has nothing close to this vastly superior technology—UFOs have been clocked flying 43,200 mph, executing instantaneous acceleration and deceleration with calculated G forces of 600, and can travel 3,315 mph underwater in our oceans! By comparison, a nuclear submarine is limited to about 30 mph underwater, because water compresses very hard like concrete at higher speeds. And more reports continue to be released.
"Estimating Flight Characteristics of Anomalous Unidentified Aerial Vehicles" by Knuth, Powell, and Reali; *USOs? Unidentified Submerged (?!) Objects* (Chris Lehto – YouTube)

The Bible doesn't explain how all angels transport themselves. Some angels fly with wings and wheels, such as the six-winged seraphim and two or four-winged cherubim who dwell with God in timeless eternity, but there is no mention of the other angels with humanlike forms having wings, contrary to many old paintings, but these angels are somehow able to fly, but it never explains how. They do need transportation to get to and from Earth, such as the angels using a ladder in Jacob's dream. The angel Gabriel said he was *caused* to fly swiftly to Daniel at the beginning of Daniel's prayer, and got there at the end of Daniel's prayer, so it took Gabriel about 3.5 minutes to physically travel to Daniel. What caused Gabriel to fly so swiftly? Could it be that some UFOs are fantastic angelic transportation crafts so technologically advanced they antiquate the spacecraft of *Star Wars* and *Star Trek*? It's unlikely angels are using 5,000-year-old human chariot technology! It is likely angels have much better technology than humans. (Zechariah 1:20; I Corinthians 15:40; Hebrews 2:7)

If UFOs do transport angels, are they good or bad angels? Could it be that there are 21 different types of heavenly species, 14 of whom are aligned under God's angel Michael, and 7 types, or a third of the stars, are ruled by Satan? (Revelation 12:3-12)

The Hebrew word "soos" means "skipping like horses and darting about like swallows" and is often translated as "horses". But in some cases "soos" is used to describe curious flying devices, such as the ones described in Zechariah called "the four spirits of heaven" who patrol Earth. The Lord's host of glowing fiery orange-red transportation devices protecting Elisha from a hostile human army were described as skipping and darting. The glowing, fiery orange-red transportation device that flew in and separated Elijah from Elisha was also described as skipping and darting, just before Elijah was taken away in a whirlwind-like vortex force. Elisha called this flying device, "the chariot of Israel and its chariot drivers", indicating that he had also seen pilots in this device. (2 Kings 2:12, 6:17) Is it just an interesting coincidence that the 1947 appearance of UFOs near Roswell, New Mexico coincided with Israel becoming a nation again? Are some UFOs the protecting "chariots and chariot drivers of Israel" from long ago that have reappeared? Their planet-wide behavior around nuclear power and weaponry supports this speculation. Details of their behavior are documented in the evidence-driven books *UFOs and Nukes* by Hastings, *In Plain Sight* by Coulthart, and *UFOs: Generals, Pilots, and Government Officials Go on the Record* by Kean.

Zechariah 1 states: In a vision at night I saw a man piloting a red transportation device that was skipping and darting over by myrtle trees that were at the bottom. Behind him were other red, bright startling red, and white transportation devices that were skipping and darting about.

I asked, "What are these, my lord?!"

The angel said, "I will show you what these are."

The man over by the myrtle trees said, "These are they who the Lord has sent to patrol Earth."

They answered the Angel of the Lord by the myrtle trees, "We journeyed all over Earth, and have found that it is at peace and resting."

Zechariah 6 states: I looked up and saw four transportation devices coming out from between two mountains. The mountains were made entirely of brass. The first transportation device skipping and darting about was red, the second transportation device skipping and darting about was black, the third transportation device skipping and darting about was white, the fourth transportation device skipping and darting about was grizzled like hail, and tan.

I asked the angel, "What are these, my lord?"

The angel said to me, "These four spirits of the air and outer space have come from the presence of the Lord of Earth who rules this place at this time. The black skipping and darting objects will go North, the white go toward after them, and the grizzled like hail will go South.

The tan came out from them and begged and beseeched and sought to go and depart and vanish to patrol Earth.

The Lord said, "Go, depart and vanish to patrol Earth."

So they departed and vanished to patrol Earth.

A literal translation of Habakkuk 3:8 and 3:15 says of the Lord's "chariot", "You pilot your skipping and darting transportation device of aid and deliverance. You push through the sea with your skipping and darting transportation device, pushing the water aside like clay cement."

Was the Lord in His misty and fiery-lit cylindrical ATD of salvation while escorting the Israelites out of Egypt, when the Lord looked down and saw the pursuing Egyptian chariots riding between the standing walls of the Red Sea, or when David later saw Him suspended in midair? Did the Angel of the Lord fly over the tops of the mulberry trees into battle before David in an ATD? David wrote that there are 20,000 ATDs piloted by thousands of angels, and the Lord is among them! (Exodus 14:24; 2 Samuel 5:24, I Chronicles 21:16; Psalms 68:17, Joel 2:1-11; Habakkuk 3:8)

Some ATDs may be piloted by wayward angels— The Bible explains the heavens are now in a great spiritual conflict. God created the thrones, dominions, principalities, and the powers who dwell there, and they are not currently visible to us, but we are visible to them, and believers are even on display as in a theatre to them. There is a great war waging among these angels, those who reside in space. Some of these celestial angels are on God's side and travel to Earth to help humans. Some angels are on Satan's side and travel to Earth to preach a false gospel. Some angels have been locked up for misbehavior. (Daniel 7:27; Acts 10:22, 12:7, 27:23; I Corinthians 4:9; Galatians 1:8-9; Ephesians 6:12; Colossians 1:16; Hebrews 1:14; 1 Peter 1:12, 3:22; 2 Peter 2:4, 2:11; Jude 6; Revelation 5:5-13, 12:7, 22:8-9)

Some people claiming contact with extraterrestrials relate that these "higher life forms" encouraged them to telepathically meditate and reach out to them for help in bringing peace and technology to humanity.

1 John 4 states we should not believe every spirit, but test the spirits to see if they are of God, by knowing if they confess that Jesus Christ is the Son of God who came in the flesh. If those on or under the earth, or in heaven or the sea, claim that Jesus was from them, or was just a very important teacher, then they are not of God. (Philippians 2:5-11)

The Angel Beast from the Abyss - Abaddon, a.k.a. Apollyon

Long ago God had the angel Abaddon locked into the Abyss, also known as the Bottomless Pit. The Hebrew "Abaddon" is usually translated as "destruction" in the five verses which give the terrible context for what Abaddon does in the place of the dead to those who rejected the Lord—

All that Abaddon does is fully exposed and open before the eyes of the Lord. Abaddon is as an unquenched fire that is never satisfied with having done enough. He burns everything of value in those dead souls who are doomed for destruction by using their iniquity that is deserving of judgment as a fire to consume their souls down to the very roots of everything that they have of value so that all that remains of their souls is an undying, withering worm. Just as our eyes are never satisfied, so Abaddon is never full or satisfied with all his consuming of lost human souls who did not believe in the Lord. God's love and faithfulness is not declared in this place of the dead, where Abaddon is. Death and Abaddon say, "We have heard a rumor with our ears of Wisdom and Understanding." But the Lord knows the way to Wisdom, and the Lord knows the place of Understanding. (Job 26:6, 28:22, 31:12; Psalm 88:11; Proverbs 15:11; Proverbs 27:20; Isaiah 66:24; Mark 9:47-49)

When this destroying angel is released from the Abyss with a key, then Abaddon, also known as Apollyon and later referred to as the Beast from the Abyss, leads an invasion of loud flying insect-like creatures with golden headgear, humanoid faces, long hair and vampire teeth who inflict terrible pain with their scorpion-like tail stingers for five months. After being cast from the heavens, Satan and his angels help Abaddon rise to power. After Abaddon's head is mortally wounded and he miraculously heals, humanity marvels and says, "Who is like him, and who is able to make war with him?" They worship Abaddon, and Satan. Abaddon is highly intelligent with amazing speaking abilities— He blasphemes God and those who dwell in heaven and wages a war to overcome believers in Christ. He gains authority in every nation and place on Earth, and his false prophet then has believers killed. Everyone whose names aren't written in the Book of Life will worship him. God warns us when this happens, that those who lead others into captivity will go into captivity, and those who fight back and kill others will also be killed. God advises us to just be patient in our faith. (Luke 10:17-20; Revelation 1:18, 9:1-11, 12:7-13, 13:1-18)

A New World Order - Deadly Superheroes

After Abaddon is released from the Abyss, God allows four angels to be unbound. They command two-hundred million riders, who kill a third of humanity. These riders have superhero-colored fiery red, hyacinth blue and sulfur yellow body armor. This is about one rider for every thirty-five humans. They ride in flying armored devices with powerful fronts which expel deadly sulfur dioxide gas and molten sulfur, and these devices have cunning, malicious tails. Near the end, Abaddon also enlists the assistance of Earth's ten new kings to demolish Earth's cities, global marketing and financial systems. About five billion humans die during this terrible time–from Satan's war on believers and humans, and from God's judgments. (Daniel 7:25-27; Revelation 6, 9:1-21, 13:1-18, 17:8,15-18)

The False Prophet

A second beast emerges from the earth, an insidiously fascinating false prophet of deceitful words and miracles, who amazes the world by calling down fire from heaven. He gains control over Earth's computers and systems. After Abaddon miraculously survives a head wound, this false prophet has humans make an image to Abaddon and then gives this image an artificial intelligence and life, or a soul, so it speaks. This artificial lifeform might arise from genetically altering injections, and/or artificial intelligence software that infiltrates humanities' smart phones and electronic devices to track down and kill all who won't worship this image of Abaddon. Only those who have Abaddon's mark, name or number on their right hand or forehead can buy or sell. Believers are then slaughtered in a global Christian genocide, and their deaths are physically avenged with the six bowls of God's wrath.
(2 Thessalonians 2:1-11, Revelation 13:11-15)

The Mark

Those who get the mark ***and*** worship the beast will be cast into the lake of fire. This "**and**", translated from the Greek "kai", means "**both**" or "**also**". "Kai" is used in all four passages about those who worship the beast ***and also*** get the mark. (Revelation 13:16, 14:9-12, 16:2, 19:20, 20:4)

What happens to those believers who get the mark of the beast, but don't worship the beast? Will they lose their hope, their guarantee of eternal life? Or, like Peter, will their receiving this mark be considered a denial of Jesus, something the Lord can use for His good in drawing them closer to Him? Could it be that some believers who get the mark to buy and sell will be like Shadrach, Meshach and Abednego, in that they participated in King Nebuchadnezzar's system of government, but refused to worship this king as a god? For those believers who do get this mark, will this deed be counted as wood, hay and stubble, and not a work of lasting value for eternity? Revelation 15:2 states that those who had victory over the beast ***and*** the image of his mark ***and*** the number of his name will stand on the sea of glass with harps of God.

Revelation 13:17-18 says wisdom is needed to understand and calculate the number of the name of this beast; it is the number of a man, 666. The Hebrew spelling of Caesar Nero adds up to 666, so this Roman emperor's AD 64 brutality against believers may serve as an analogy of how Abaddon and his false prophet will wage their war against believers. (Hebrew letters also stand for numbers. For example, the Hebrew "6" is the sixth Hebrew letter "Vav", which was original written as a "W" in the Hebrew. So, "666" is the old Hebrew "WWW".)

First Four Seals and the Lord's Four Spirit Horses, and the Trumpets

Zechariah 1 provides the backstory for the four horses of Revelation 6. The riders of the red, white, and brown horses (or rather angelic pilots if these "horses" are angelic transportation devices) are in some myrtle trees after returning from a patrol of Earth. The riders tell the angel of the Lord that Earth is resting peacefully. In Zachariah 6 these horses, and a black horse, are sent out for the Lord's work. Revelation 6 completes this analogy of the four horses—

The **White Horse** and crowned rider Goes Forth among the nations and brings victorious conquering— This represents Christ's victory out of Israel over the past few thousand years. Turkey, Germany, England, Spain and the Americas have historically been predominantly Christian. Libya and Algeria were primarily Christian up to the seventh century, before the Muslim religion began to dominate these areas, and some of the other areas as well. (2 Timothy 4:8; 1 John 5:4; Revelation 19:11-12)

The **Red Horse** is Everywhere and takes peace from Earth and there is war and slaughter— In Zachariah 1 the rider of the red horse is the Lord's angelic messenger who reports back to the Lord of Heaven's armies. In Revelation 6 the red horse takes peace from Earth and there is war and slaughter everywhere.

The **Black Horse** goes North and there are high food prices— The black horse was not mentioned in Zachariah 1 when Earth was resting peacefully. In Zachariah 6 it goes North to give rest to the Lord's Spirit, where most of Earth's food now originates in the vast grain fields of Ukraine and Russia. Revelation 6 reveals the black horse is associated with high food prices. A reduction of grain farming in this land north of Israel will cause worldwide food shortages and high food prices, since chicken and beef are derived in part from these grains as well.

The **Pale Green Horse** goes South and is given authority over 25% of the earth to kill with widespread famine, medical shortages, disease, and wild beast(s)— The brown horse patrolling the peaceful earth for the Lord in Zechariah 1 becomes grizzled and bay horses in Zechariah 6, and the bay horse is eager to patrol Earth again and is sent out. The Hebrew for "grizzled" means spotted, as if spotted by hail. This corresponds to the fire and hail, which may be meteorites, that cause a third of Earth's trees and grass to burn after the first trumpet of Revelation 8:7. It becomes a pale green horse in Revelation 6, which may represent this wilting and burning of the green grass and trees. If, for example, one third of Earth is hit by the meteorite blast south of Israel, then Africa, Brazil, and fringes of Europe and the eastern part of the United States will be affected. It's unclear if the brown horse is changing to grizzled and bay and then pale green, or if they are all separate horses.

The rider's name of this pale green horse is "Death", and "Hades", the place of condemned souls, follows after him. This associates the rider of the pale green horse with Abaddon, the destroying angel beast from Hades. There is a killing "beast" that is with this pale green horse, it is translated from the Greek "therion" and used later in Revelation 17:8 for the "beast" from the Abyss, which is the angel Abaddon—

In Revelation 9:1-2, a star falls to the Earth from space and is given the key to the shaft of the Abyss. This star then releases the angel Abaddon, who is the beast (therion), from the Abyss of Hades. This indicates that Death, the rider of this pale green horse, may play a part in giving the key for the shaft of the Abyss to this star (stars are off-planet beings not from Earth). Abaddon may emerge from this portal abyss in Africa, since the pale green horse goes south and is given authority there over one fourth of the earth. (Revelation 1:18, 20:14)

The fifth seal of martyred saints, and the sixth seal with the great earthquake in Revelation 6:9-17, occurs after the beast and the false prophet are on Earth. (Revelation 12:17, 13:15, 16:18, 20:4)

Overlapping Revelation Events in Chronological Order					
1st SEAL White Horse - Babylon the Great					War in Heaven: Michael v. Satan Rev 12:7-12
	2nd SEAL Red Horse - Peace is taken from Earth				The 7 Trumpets Rev 8
		3rd SEAL Black Horse - Scarcity of food			1. Meteorites - 1/3 of land burns
					2. Mount of Fire - 1/3 of sea burns
					3. Wormwood - 1/3 of fresh water is polluted
					4. Earth's Sky is Blocked - 1/3 of sky is blocked
			4th SEAL Pale Green Horse - Death with Hades	5th SEAL - Martyrdom Deaths and the 2 Witnesses	5. Abaddon and Locusts - Abyss opens Rev 9
					6. 4 Angels unbound and 200 million pilots
					Abaddon the Beast begins his rule of Earth
					The False Prophet's Miracles Rev 13
					Abaddon's head wound heals
					The Mark - forced worship of the Beast
					*1st Harvest of all Beleivers Rev 14:14-16, 20:4
					7th Trumpet and 7 Bowls of Wrath Rev 11,16
					White Clouds- Jesus approaches Earth Mat 24:30
					The 3 Evil Spirits gather army Rev 16
					The Army of the Beast and the 10 kings Rev 17
				6th SEAL - Earthquake	Earthquake levels the entire planet Rev 16
					The mountains and islands are no more
					Jesus and the saints are revealed in space Rev 19
					Jesus conquers the Army of the Beast
					The Beast and False Prophet cast in Lake of Fire
					Satan is cast into the Abyss

*An angel comes from the temple of God the Father and tells Jesus it is finally the time for the "First Harvest", the "catching away of the saints".
(Matthew 24:31-36, I Cor 15:52, Revelation 14:14-16, 16:15)

<u>Subsequent Events after the 7 Trumpets and 7 Bowls of Wrath</u>

Millenial reign with Jesus for 1000 years Rev 20

Satan is then released from the Abyss and rebels with the nations

Satan is cast into the Lake of Fire

Old Earth burns with fire

White Throne Judgment

1. The 2nd Harvest of unbeleivers - Second Death Rev 14:17-20
2. The deeds of beleivers are judged and rewarded I Corinthians 3:11-15

Death and Hades are cast into the Lake of Fire

New Earth - All things are made new Rev 21

New Jerusalem - God's City decends

God's Throne, River, Street, and the Fruit Tree of Life Rev 22

The Rise

Is there to be a rapture in which Christians who are still alive on Earth suddenly vanish and leave behind a pile of clothes before a great tribulation? Many fictional books on eschatology, the study of biblical end times, have portrayed a vanishing rapture *before* the great tribulation, such as *In the Twinkling of an Eye*, published in 1911. The origin of this belief, of Christians suddenly disappearing before a seven-year tribulation, is attributed by some to John Darby back in the 1800's. The 1909 *Scofield Reference Bible* helped popularized this idea of a rapture, which became a strong belief for many since then. Clarence Larkin, whose charts from his 1918 book *Dispensational Truth* influenced conservative churches, reinforced this as a doctrine. *The Late Great Planet Earth*, published in 1973, proclaimed this rapture would occur before 1990. Then came the *Left Behind* series of the 1990's. These books have long been instrumental in leading many to believe in Jesus, which is a good thing. The problem is, this idea of a rapture before the tribulation is based on an opinion, or an interpretation, of a few verses, and could very well be in error. Could there be a rapture before the tribulation? I've thought so in the past. I read *In the Twinkling of an Eye* in my teens, had a *Scofield Reference Bible*, was raised in a conservative Christian group that used some charts from *Dispensational Truth*, read *The Late Great Planet Earth*, and watched a movie based on *Left Behind.*

This idea of a "rapture" is based in part on 1 Corinthians 15:52, which reveals that the bodies of mortal believers will put on immortality in the "twinkling of an eye" at the "last trump". While I was investigating the relevant passages and the book of Revelation, it slowly dawned on me that I was unfortunately wrong about a pre-tribulation "rapture". My investigation found that this physical putting on of immortality in the "twinkling of an eye", when believers are instantaneously transformed into their new physical superbodies, is likely just before Jesus descends to Earth to wrest it back from Satan. This "last trump" is likely the last seventh trump near the end of the tribulation in Revelation 11:15. I tried very hard, but simply could not prove this twinkling of an eye event occurs *before* the tribulation! ☹

The Day of Christ, when Jesus comes as a thief and we are gathered to Him, is *after* the falling away, and *after* the false prophet is revealed as the man of sin, the son of perdition who exalts himself above all gods. It says to let no one deceive you otherwise. I suspect Christianity will be discredited if a "rapture" does not occur before the tribulation begins, and if so, I hope believers are not deceived and fall away from their faith. (2 Thessalonians 2:1-11)

Jesus said in Matthew 24:36 that even He doesn't know when we will be gathered to Him, and only God the Father knows. Revelation 14:14-16 may tell of when God the Father finally lets Jesus know it is time for this "twinkling of an eye" event— An angel comes from the temple of God the Father and tells one wearing a golden crown who looks like the Son of Man that it is time for Earth's first harvest. Jesus then sends His angels to gather His believers up in a flash, and, as it seems, we are beamed up by their fiery ATDs, just as Elijah was. We are instantaneously transformed into our new quantum-enabled multi-dimensional physical superbodies as we rise to meet the Lord in His great cloud, soon before He descends down to Earth to take it back from Satan. Those who already died, whose souls are already with the Lord, will receive their new superbodies first, then those believers who are still living on Earth will be transformed into theirs. (2 Kings 2:11-12; Daniel 7:9-10; Matthew 24:31,42-43; 1 Thessalonians 4:17; Revelation 4:2-5, 6:4, 13:1-17, 19:15)

The Date

Even though Jesus said only God the Father knows when this tribulation will happen, and just told us of the signs we should look for, some have still made predictions... Alex Haley, who wrote, *The Late Great Planet Earth*, erroneous predicted a rapture would take place in the 1980's. He based this prediction on Matthew 24:32-34—

Jesus, when privately telling His disciples about the tribulation, said, "This generation will not pass from the scene until these things take place." There has been a longstanding speculation that this generation began when Israel became a nation in 1949. However, if this idea is true, which it could be, the Bible doesn't spell out exactly how long this generation is. For example, God says in Genesis 6:3 a lifespan is 120 years, Psalm 90:10 indicates 70 or 80 years. Genesis 15:16 says that the Israelites would be in Egypt for 4 generations, or 400 years, but they were actually there for 10 generations, which is 40 years per generation. So, is a generation 40, 70, 80, 100, or 120 years? It's unclear. Alex Haley figured a generation was 40 years and thought the rapture had to be before 1990.

The Millennial Day Theory

The Millennial Day Theory speculates that the seven days of Genesis represent a 7,000-year period for modern humanity. If each "day" of Genesis represents a 1,000-year "millennial day", then the seven days of Genesis may represent seven millennial days. The Millennial Day Theory is based in part on 2 Peter 3:8 and Psalms 90:4, which state one thousand years is like one day for God— The first six millennial days of work culminates with the Tribulation, and Christ's millennial reign at the end corresponds to God's seventh day of rest, for a total of 7,000 years.

The first millennial day begins with Adam's exit from Eden around 4000 BC. The crucifixion of Jesus marks the end of the fourth millennial day, the end of the first 4,000 years. 2,000 years after His crucifixion places the end of this 6,000-year period in *about* AD 2033. (Our current calendar system counts the years from His *birth* until the return of Jesus, but if we instead counted the years from His *crucifixion*, then our current year AD 2033 would actually be *about* the year 2000)

Most bibles date Adam to around 4000 BC. They use the genealogical dating of Jerome's AD 405 Latin Vulgate translation from Pharisaic text that was standardized after the AD 70 temple fire. However, Augustine's genealogical dating from the earlier 200 BC Septuagint LXX dates Adam to about 5500 BC, and ends this 6,000-year period in about AD 500.[17]

Regardless of when modern human civilization ends, it will be obvious the tribulation is occurring with these R-rated events— Fire burns one third of Earth, Abaddon emerges from a portal with abominable creatures who torture humanity for 5 months, two-hundred million extraterrestrials invade Earth and two thirds of humanity are killed, two witnesses from the Lord kill anyone who attacks them, anyone who won't worship Abaddon is killed, then the global economy and structures are destroyed, then six bowls of God's wrath are poured out on Earth (similar to the Egyptian plagues with Moses), then the sign of Jesus coming from deep space, then the gathering of a great army to fight Jesus and His army approaching Earth in a great cloud, then God's planetary earthquake levels the mountains and islands, and finally Abaddon's great army is quickly vaporized by Jesus and His army. The amazing 1,000-year reign of Jesus Christ begins soon afterwards. Earth will peacefully flourish with long-life spans and great social and technological advancements under the rule of the Lord of the Sabbath. (Matthew 12:8; Acts 1:7; Revelation 20)

The Heptad Civilization Theory

What if the speculated seven-thousand-year civilization period for modern humanity began soon after Noah's flood, since Adam's era ended with God's judgment on the cruel violence? The Septuagint LXX genealogy aligns Noah's flood with historic floods near Ur and Uruk about 3300 BC, where Earth's first city dynasty governments and corporations emerged soon afterwards during their Jemdet Nasr period, when human history was first written down, on clay cuneiform writing tablets.[17]

It makes for some interesting analogies to compare the six sequences of Genesis 1 with the recorded history of human achievements—

Age 1 3000 BC Enlightenment: Human knowledge enlightens the pre-historic darkness with Mesopotamian and Egyptian writings, and modern civilization's political, architecture, and agricultural skills emerge.

Age 2 2000 BC Chaos to Order: Human firmaments of law and order form in Mesopotamia, Egypt, Indus Valley, Levant, Andean, and Chinese societies, and theologies separate the divine from the human realm.

Age 3 1000 BC Growth on Land: Intellectual plants of human philosophies, laws, and politics thrive and grow in the lands of Israel, Greece, Rome, Persia, and South America.

Age 4 AD 1 Sun, Moon and Stars: World religions and intellectual traditions shine with the Jewish (star of David) and Islamic (moon crescent and star) religions, and with Confucianism, Buddhism and Greek philosophies. Jesus is born, understanding enlightens, and the calendar is established.

Age 5 AD 1000 Sea and Air: Vikings and global cargo ships navigate the seas, and skyscrapers, planes and satellites fill the air. Humans migrate by sea and air amidst medical, scientific, and industrial progress.

Age 6 AD 2000 Mammals and Humans: Humans dominate Earth with their inventions and create intelligence in their own image.

Age 7 AD 3000 Millennial Rest: Christ is Earth's King, and Earth rests.

Of course, this is just another theory, but it helps demonstrate that the end time events of Revelation could begin this year, or in a thousand years. Who knows? Jesus said only God the Father knows the future. But, in the meantime, He calls us to be witnesses for Him. (Acts 1:7-8)

Is the Tribulation 7 or 3.5 Years?

Daniel 9:24-27 indicates how long the tribulation is. The backstory on this passage is that Israel had sinned for a long time, so God allowed the Babylonians to begin trampling on Jerusalem in 606 BC when Daniel was taken captive. They later destroyed Jerusalem in 586 BC, and more Jews were forced to immigrate to Babylon. These Jews yearned for their home country, to rebuild the walls of Jerusalem and their temple, but the Jews were *especially* waiting for their Messiah, who was to become their King and make all things right for them. (Isaiah 11:1-12)

In Daniel 9:24-27, Gabriel reveals to Daniel that, beginning with the command to restore Jerusalem, 490 years are determined to anoint the most Holy, when the Messiah will finally become their King. (Keep in mind a time gap kicked in after Jesus was rejected as their Jewish king and died on the cross, so this prophesy is now suspended and has been put on hold during our current time of the church.) The question is, how much of this 490-year prophesy was used up and how much of it is still left? Knowing how much time is left of this prophesy tells us how long the tribulation will be, because Jesus will become the King at the end of the tribulation. Is the tribulation going to last for 7 or 3.5 years?

Gabriel said first Messiah the **Prince** would come at 69 weeks (each day of a prophetic week is a year, so 69 x 7 = 483 years). Messiah the **Prince** was Jesus who came right at 483 years, and confirmed a covenant with many, but in the middle of this last week of 7 years Jesus was cut off, not for Himself but to end all sacrifices and offerings (Jesus made all sacrificing obsolete per Hebrews 9-10 when He offered Himself as the ultimate sacrifice, God dying for humankind). Gabriel said after Jesus is cut off, then the tribulation will begin "on the wing of abominations will be one who makes desolate (Revelation 9:1-11), until the consummation is poured out on the desolate (Zechariah 14:12; Revelation 19:21)." Messiah the **Prince** arrived at 483 years, but about 3.5 years later, when He entered Jerusalem to become Messiah the **King**, he was rejected and cut off for the sins of humanity (Isaiah 53:7-8, Daniel 9:26). So, **486.5** years were used up and **3.5** years are still left of this **490-year** prophesy! For the last 3.5 years, the saints will be given into the hand of the beast: Wars, violence, famine and death may precede the beast, but the great tribulation period with its extraterrestrial events seems to be 3.5 years, not 7.

(Daniel 7:25,12:7; John 12:13, 18:37; Revelation 9, 12:14, 19:16)

The Prophesy of Daniel 9:24-27 - in Gabriel's Words

Weeks 70 are determined upon your people and holy city to finish transgression and end sin and make reconciliation for iniquity and bring in everlasting righteousness and seal up vision and prophecy and anoint the Most Holy. From the command to restore and build Jerusalem until Messiah the Prince weeks will be weeks 7 and weeks 62. Again will be built the street and the wall in troublous times.

(The first 49 years *were* troubled times per Ezra and Nehemiah.)

After weeks 62 will be cut off the Messiah but not for Himself.

(Jesus was cut off **after** week 69 and the prophesy clock froze.)

The city sanctuary is destroyed.

(The temple was destroyed in AD 70.)

The flock of the Prince abides until the end comes.

(Jesus is the Prince of Peace per Isaiah 9:5, now sits at God's right hand per Hebrews 8:1, is the Prince of Life per Acts 3:15, and returns as Earth's King per Revelation 19:16. Believers are the flock of Jesus per Matthew 26:31. Satan is the prince of this world and power of the air and does not have a flock. The beast, the false prophet and man of sin are never alluded to as "prince" and do not have a flock.)

Flood until the end: War is determined and desolations.

(Deluges of wars until the end, when planet-wide desolations occur)

Last Week - The Two Events Before and After the Cross

For 3.5 Years Before the Cross: Confirms a covenant with many week one. In the middle of the week ceases sacrifices and oblation.

(In the middle of the last week, after His 3.5-year ministry was complete, Jesus confirmed the new covenant at the last supper per Matthew 26:28, and ended sacrifices on the cross per Hebrews 10:12. There is no New or Old Testament context for an evil "covenant". Daniel 11:22 references another prince of the covenant— this was likely Onias the high priest who was ousted by Antiochus Epiphanes.)

After the Cross for 3.5 Years: Against the wings of abominations desolation until the consummation determined is poured on the desolate.

(Abaddon's arrival with winged abominations inflict desolations per Revelation 9 is what seems to re-start the prophesy clock. The bowls of God's wrath, and the return of Jesus as King, is the consummation "poured" on the desolate at the end of the tribulation.)

<table>
<tr><th colspan="6">Weeks 70 to Make Reconciliation for Iniquity and To Anoint the Most Holy - Daniel 9:24-27</th></tr>
<tr><td>First 7 Weeks (49 years)
Begins in 445 B.C.
Jerusalem is Rebuilt</td><td colspan="4">Next 62 Weeks (434 prophetic years)
To 30 A.D.
Waiting until Messiah the Prince</td><td>Last Week (7 Years)
70th Week is from
30 A.D. to ?</td></tr>
<tr><td colspan="6"></td></tr>
<tr><td colspan="6">70th Week - 30 A.D. to Jesus Arriving as King of Earth</td></tr>
<tr><td colspan="2">First Half of Week 70 - 30 A.D. to 33 A.D.</td><td rowspan="2">Cross
Pauses 70
Week
Prophesy
(This Age)</td><td colspan="3">Last Half of Week 70 - Dates TBD</td></tr>
<tr><td>Jesus the Prince
Arrives after week 69
and the Beginning of
week 70</td><td>Jesus Confirms the
Covenant, is Cut Off,
and Ends Sacrifices</td><td>Abaddon
Arrives</td><td>Consummation is
Poured on the
Desolate</td><td>Jesus the King
Arrives at the End
of the 70th Week</td></tr>
<tr><td colspan="6"></td></tr>
<tr><td colspan="6">Prophesy Pauses - Events From 33 A.D. to an unknown Future Date</td></tr>
<tr><td colspan="6">The Temple is Destroyed in 70 A.D.</td></tr>
<tr><td colspan="6">Flock of the Prince Comes - From 33 A.D. until Near The End</td></tr>
<tr><td colspan="6">Overflowing until The End - War is Determined and Desolations</td></tr>
</table>

Daniel's AD 30 Prophesy for Jesus

Gabriel told Daniel when Jesus would come as the **Prince**—

"Know and understand that seven weeks plus sixty-two weeks will pass from the time the command is given to rebuild Jerusalem until Messiah the **Prince** comes." (Daniel 9:24-25, Matthew 4:17)

This time span is (7 x 7) + (62 x 7) = 483 years.

This time span begins with the command to rebuild Jerusalem, given by the King Artaxerxes in the first Jewish month of Nisan in 445 BC. This time span ends with the coming of Jesus as Messiah the **Prince**. (Nehemiah 2:1; Matthew 4:16-17; Mark 1:15; Luke 4:18-19; John 1:36)

These prophetic concepts of time need to be understood:
Prophetic time is one year for each day.
Prophetic time is multiplied by seven for punishment.
The Jewish calendar year is 360 days, with twelve 30-day months.
(Genesis 4:15,7:11-24,8:3-4; Leviticus 26:18,24,28; Numbers 14:34; Ezekiel 4:6; Revelation 11:2-3, 12:6-14, 13:5)

Beginning with 445 BC, there were 483 prophetic years until Jesus was to come as the Prince. However, this places His coming on the Jewish calendar, not our Gregorian astronomical calendar.

The Gentile (non-Jewish) astronomical year has 365.25 days. Since it skips from 1 BC to AD 1 and has no year zero, time spans that cross over from 1 BC to AD 1 need to have this "Year Zero" subtracted.

To convert 483 prophetic Jewish years to our current calendar
483 x 360 / 365.25 = 476.06 astronomical years, or 476 years and 21 days from the command to rebuild that was given in 445 BC.
Starting at 445 BC, and subtracting for Year Zero, brings this to AD 30 when Jesus began to preach as the **Prince**. The exact year is not historically confirmed, but most estimates are close to AD 30.

Jesus had rebuked the Jewish rulers for being a religious "brood of vipers," so they didn't care how many people Jesus had healed, blind made to see, or paralyzed enabled to walk. These self-righteous Jewish religious rulers influenced the Jews to reject Jesus as their **King** about three and a half years later, around AD 33.
(Daniel 9:26-27; Zachariah 9:9; Matthew 21:9-11, 27:20; John 12:12-15, 18:37)

Anti-Semitism

The Jews were despised after Jesus was crucified, after saying, "Let His blood be on us and on our children." (Matthew 27:25)

Rome: In AD 50 30,000 Jews were killed, and in AD 70 about one million were killed or starved by Titus, who later disclaimed responsibility because it was so horrific. In AD 135 Hadrian killed 580,000 Jews, and they were persecuted long after Constantine's Edict of Milan in AD 313.

England: In AD 1096 all Jews were killed who would not be baptized. In AD 1190 Jews were murdered and their homes plundered. English Royalty claimed to own Jews and their possessions, and in AD 1290 they were expelled from England.

France: In AD 1305 100,000 Jews were expelled and their possessions confiscated. When they returned, Pastoureaux shepherds drove them away again. In AD 1683 Jews were ordered to leave all French Colonies.

Germany: In AD 1349 the Black Death hadn't affected Jews as much, due to their clean hygiene standards arising from their Levitical Law. The Flagellants accused them of causing this disease and tried to exterminate them— 2,000 Jews were burned together alive in Strasburg. In AD 1560 they were banished from Germany after 3,000 Jewish houses were burned.

Spain: By 1492 the Spanish Inquisition had killed many Jews, and imprisoned others in dungeons. The Iron Maiden was one infamous casket-of-iron torture device; as "her" door closed, blunt rusted knives pierced the shrieking victim's bodies, and they were dropped into a pit below. In AD 1492, Ferdinand and Isabella banished all Jews from Spain. Jews came before them and offered lots of money to revoke this, but the Grand Inquisitor Torquemada came in, holding a crucifix, and accused them of selling out Jesus as Judas Iscariot had. The King and Queen caved under this religious intimidation, and the Jews had nowhere else to go, having already being banned from England, France and Germany.

The tide turned in AD 1753 with England's Naturalization Bill, in AD 1776 with the Declaration of Independence, and in AD 1790 when some Jews were granted full citizenship in France. They survived as a people without a country, even the Holocaust that claimed two-thirds of their European population by AD 1945. (Zechariah 13:7-9)

Ironically, the persecuted Jews helped expose these barbaric nations to God's superior moral and conduct codes, and civilly influenced them with Old Testament biblical laws, wisdom and proverbs.[18]

Jesus Predestines AD 1949 for the Nation of Israel

In Luke 21:24, Jesus said, "Jerusalem will be trampled down by the Gentiles until the time of the Gentiles comes to an end."

It was speculated for centuries that this time refers to a 360-day prophetic year, with each day lasting one year, and, like the 70 weeks of Daniel, were multiplied by seven for punishment. Were the Jews to be without a country for 360 prophetic years x 7 = 2,520 years?

This trampling down began when Daniel and other Jews were first deported from Jerusalem to Babylon in 606 BC, some 20 years before Jerusalem's destruction in 586 BC. (Daniel 1:1-7; 2 Kings 24:1; 2 Chronicles 36:5-6; Matthew 1:11; Babylonian astronomical diary VAT 4956)

Converting 2,520 Jewish 360-day years to 365.25 astronomical day years yields AD 1876. But Israel did not become a nation in AD 1876, so this speculation faded. (360 x 7 = 2,520 x 360/365.25 – 1 – 606=1876)

However, since this is the time of the Gentiles, should a Gentile (non-Jewish) 365.25-day astronomical year be used instead, since Daniel and Jeremiah switched to Babylonian king accession year dating while under their rule? (The year *after* the king's actual accession was counted as the first by the Babylonians, but as the second by the Jews and Egyptians).

> 365.25 x 7 = 2,556.75 years – Year Zero = 2,555.75 years.
> Counting 2,555.75 years from 606 BC ends in AD 949.
> The United Nations officially acknowledged Israel in 1949.
> The US recognized them in 1948, and Great Britain in 1950.

Jesus *does* seem to indicate when Israel would again be a nation per Luke 21:24! The prophecies of Jews coming from all lands, no longer as twelve tribes but as a homogenized nation of people in their own country, has certainly come true! (Ezekiel 22:19-22, 20:34, 37:12; Jeremiah 16:14-15)

Hosea 3:4 states when Israel is once again a nation, they "shall abide many days without king or prince, without sacrifice or sacred pillar, without ephod or teraphim"— This perfectly describes Israel in their current state, without a king or a temple or sacrifices.

If the time of the Gentiles ended in about AD 1949, when will the Great Tribulation begin? Jesus said only God the Father knows this, so any future date set by a human is doomed to fail.
(Daniel 9:27; Matthew 24:32-36; Acts 1:7; Revelation 14:15)

The Woman of Israel

God designed us to desire marriage for fulfillment. How a man feels about the woman he loves is how God feels about a nation. The Lord grieves over the loss of relationships with nations of people like a man grieves over a wife he loves who is prostituting herself. (Hosea)

After Israel twisted what is right with fraud and violence, and even sacrificed their own children in fire, the Lord said, "You defiled your beauty on every street corner and offered yourself to everyone in an endless stream of prostitution and promiscuity. They were shocked by your lewd conduct and insatiable lust for new lovers! You have a sick heart to act like such a shameless prostitute. You build your pagan shrines on every street corner and your altars to idols in every square. In fact, you are worse than a prostitute– you're so eager you don't even want to get paid! Prostitutes charge for their services, but you give your lovers gifts to bribe them into having sex with you! So, you are the opposite of other prostitutes, you pay your lovers instead of getting paid! You have played the slut with many lovers; Yet, return to me."

The Church Bride

Christ loves all believers like a husband who greatly loves his wife. Husbands should take care of their wife as if they are caring for themselves. A man who loves his wife is loving himself. This is how Christ loves the church. Jesus gave His life for the church, which is composed of all believers, and washed her clean with God's word because He wants her to be without any blemish; holy and faultless. Just as a man is joined to his wife and they become one, so Christ and the church are one. Believers are members of His body, and as in a marriage, are joined to the Lord in one spirit. To the Lord, when believers are immoral, it is like a wife who prostitutes. It grieves Him.

The Prostitute

At the end, during the tribulation, a vast global system rules over all the nations and Earth's masses of humanity. This system is used to buy, sell and ship great quantities of fashionable clothing, jewelry, all kinds of food, merchandise, products and vehicles. The nations love this great global system like a prostitute, and are intoxicated with her wine of passionate immorality. She uses slaves and buys the souls of humans. Earth's rulers commit adultery with her, because of her desires for extravagant luxury, and Earth's merchants become fantastically wealthy.

She becomes a home for demons and a hideout for foul spirits, and deceives the nations with her sorceries, drugs and pharmaceuticals. She gets drunk with the blood of believers after slaughtering them in a worldwide genocide, and glorifies her evil deeds and cup of terror as she lives on in luxury. She says, "I have no reason to mourn! I am queen on a throne, not like some helpless widow."

But in a single day this vast global system is consumed by fire— Near the end, ten rulers are appointed to ten kingdoms for a brief time. Because God puts a plan into their minds to carry out His purpose and judgment, they unite and give Abaddon their power and authority. These ten rulers and Abaddon hate this prostitute, strip her bare, consume her, and burn what's left. Those who enjoyed her great luxury are terrified by her torment and mourn as they see the distant smoke rising from her charred remains. They cry out, "All your luxuries and splendor are gone, the luxuries we love so much are gone forever!"

The merchants and shipowners who became wealthy by transporting her great wealth across the oceans weep and mourn because no one is left to buy their goods. The captains, crews and passengers of cargo and cruise ships cry out as they watch her distant smoke ascending, and say, "Where is another as great as this? How terrible! How terrible this is! She was so beautiful, and in a single moment all of her wealth is gone!"

A mighty angel then shouts, "Musicians will never be heard in you again! No craftsmen and no trades will ever be found in you again! The sound of manufacturing will never be heard in you again and your lights will never shine! For your merchants were the greatest in the world, and you deceived the world with your sorceries. In your streets flowed the blood of God's people, and of people slaughtered all over the world."

A vast crowd in heaven then shouts, "Praise the Lord! For our Lord God the Almighty reigns. Salvation and glory and power belong to our God. His judgments are true and just. He judged the great prostitute who corrupted the earth with her immorality for the sake of His believers, and avenged the murder of His servants. Let us be glad and rejoice, and give Him honor because the time has come for the wedding feast of the Lamb, and His bride has prepared herself. She has been given the finest of pure white linen. Blessed are those who are invited to the wedding feast of the Lamb." (Isaiah 62:5; Jeremiah 3:1; Ezekiel 16; 23; Hosea 1-3; John 3:29; 1 Corinthians 6:15-17; Ephesians 5:25-33; Revelation 17:7-19:9)

Immortal Superbodies

Job said, "I know my Redeemer lives. He will stand on Earth in the end after my body has decayed, yet in my body I will see God! I will see Him for myself. Yes, I will see Him with my own eyes!" When they hear of it, some intellectual people make fun of the dead being physically resurrected, but others want to know more—

We are not at home with the Lord as long as our biological shells are alive, but after death our soul and spirit goes to be with the Lord. Will our soul and spirit still be able to see, hear and remember? Samuel's spirit could see, hear, and think when he arose from Paradise to talk with Saul. When Paul was caught up to the third heaven he couldn't tell if he was in or out of his body, but he could still see! In Revelation, when John's spirit entered God's timeless dimension, he could still see, hear and recall details about God's throne without the aid of his physical body.

So, why do we need a physical body? This is the main point that Jesus makes about what He wants to do for us! The Lord told Adam long ago that if he disobeyed, then, "You will surely die." Adam could have biologically lived forever by eating from the tree of life, but a flaming sword guarded it after Adam sinned. So, God foreordained Jesus to provide the way for us to live eternally in new, physically superior bodies.

Jesus said, "**I am the resurrection and the life**: Anyone who believes in me will live, even after they die. Everyone who lives in me and believes in Me will never die." Our soul and spirit live on. But this isn't enough for God, He wants us to live in upgraded, and vastly superior, new physical bodies, as superior as the sun is to the moon. Paul compared our mortal, aging, disease-prone earthly bodies to a fragile tent that soon perishes, but our new eternal bodies will be an immortal permanent home for our soul and spirit that endures forever.

God wants us to live on forever, not merely with just our spirit, but with amazing new physical bodies! Think superhuman abilities. Amazing ones. Jesus walked through walls, teleported, rose in the air, yet also ate fish and honeycomb after His resurrection. He still looked the same. He still had evidence of his injuries. Jesus showed his disciples that He was not a spirit, but was Himself with a flesh and bone physical body. *Everyone* agrees that the body of Jesus was missing after the crucifixion, it was because God remade His physical body and brought Him back to life!

Believers are now spiritual children of God, and we will someday be physical children of God after He refabricates our new bodies. He will make us equal to the angels who never die, so we won't reproduce. Since our now inferior biological bodies simply can't last forever, no matter what genetic enhancements are made, they *have* to be transformed into immortal ones so we can physically, and eternally, inherit the Kingdom of God. After our bodies are transformed, this Scripture will be fulfilled— "Death is swallowed up in victory. O death, where is your victory? O death, where is your sting?"

Will our bodies be refabricated from our decayed flesh? What of those who were cremated? What of those who decomposed in a corn field long ago, and their atoms later became part of another's body? How will God decide who gets which atom? I suspect ordinary atoms as we know them won't be involved, nor will our decomposed bodies. Just as God didn't explain quantum physics to us, He isn't explaining this mystery either. Perhaps someday He will explain both to us.

Jesus made it simple— Believers will receive new, upgraded physical bodies to live with Him forever on an upgraded and new physical earth.

We *are* God's children, but God hasn't told us what our new bodies will be like. He just let us know that when Jesus appears, we will be like Him, because we will see Jesus as He actually is. God will give to each one of us the new spiritual body that He wants us to have. Jesus said that this is God's will, and why God sent Jesus, so that all who believe in Him can live forever in new, superhuman bodies that never wear out.

God gave us a hint of the future just after Jesus died— The graves were opened and many bodies of the saints which **slept** arose from their graves, and then they went into the holy city and appeared to many people. This is very interesting, because it explains that <u>those who sleep</u> refers to <u>the physical bodies of those believers who have died</u>.

After Jesus was physically resurrected, his followers watched Jesus rise and depart from their sight into a cloud. As they stood watching this cloud, two men standing nearby in white clothes said, "Why do you stand there looking up at the cloud? Jesus has been taken up, but He will come back to Earth just as you saw Him leave."

Jesus assured us this *will* happen, and said, "You will see me coming in the clouds of heaven."

(See the next section for references)

The Return of the Lord

The descent of Jesus towards planet Earth will be unexpected, but it becomes more obvious that a major event is about to occur after four angels hold back the four (magnetic?) winds of Earth.[19,20,21] Our solar system becomes disturbed and celestial bodies begin falling to Earth. A third of Earth's vegetation burns, a third of Earth's water becomes contaminated, and a third of Earth's sky becomes obstructed.

Sometime after this, Jesus approaches Earth with His armies of angels from space with magnificent power in transportation of some type, but all the people of Earth can see of them are the great clouds concealing them. They mourn as Jesus and his army approaches Earth in this mysterious cloud. The sight is tremendous. It is obvious a major event is inevitable. With the Lord are the saints and believers who have died.

Jesus said only God the Father knows when this will be. When the unbelievers of Earth are saying, "Everything is so peaceful and secure," then disaster will suddenly fall upon them and they will have no escape. Jesus will unexpectantly raid Satan's earth domain to steal away His believers' physical bodies like a thief in the night. People will be at nightclubs, watching movies or be at weddings, and won't even realize this global event is about to happen. Two people will be working in construction or in an office, or two will be in bed, depending on their time zone, and then one will suddenly be taken and the other left.

An angel will come from God's temple to say it is time. Then, with a commanding shout and the voice of an archangel, when the last trumpet is blown God will call His chosen ones. Powerful angels will then gather us up as a harvest from all over Earth and space—

The physical bodies of believers who "sleep" will rise first from their graves, and each believer's soul and spirit will receive their new, and vastly superior, refabricated body. Job will finally be able to physically stand on Earth and see the Lord with his own eyes!

The Lord said no one can see His face and live. The Lord covered the face of Moses with His hand as He walked by Moses on Mount Sinai. It may be that we will need these upgraded bodies so that it won't kill us when we see the Lord's face (after Moses was in the Lord's *presence*, his face radiated with so much light that it frightened everyone).

After the believers who already died have received their new bodies, *then* the physical bodies of all the believers who are still left alive on Earth will be physically transformed in the blink of an eye, and they too will be caught up into these clouds to meet the Lord. Then, all believers will be with the Lord forever in their new, physical superbodies. But, we will all be up in this great cloud with the Lord, and not yet here on Earth.

Satan and his angles have worked through and with Abaddon and his false prophet, utilizing Earth's governments, computers and systems, to slaughter millions of believers, and God sends horrible "bowls" of plagues to avenge their deaths. When the sign of Jesus approaching Earth is seen, they send out three unclean spirits, like frogs, to gather and unite the ten kings of Earth and their armies to fight this invader from space—even though the Lord made them and created Earth.

The clouds roll back to reveal Jesus and His armies, and they descend to fight in Earth's tremendous battle. From the perspective of humans two thousand years ago they are on clean, pure white horses, but from the perspective of science fiction fans they are in gleaming white spaceships. If other white spaceships appear before this on Earth, or rise from the sea, or if anyone performs miracles and claim to be Jesus, it will not be Jesus. Jesus will come from space in a great cloud. After an earthquake levels Earth's mountains and islands, Jesus and His army fly over in perfect formation and the energy force emanating from the front of their ATDs consumes Satan's army so rapidly that their living flesh-their bodies, eyes and tongues, wither away as they stand.

Abaddon and his false prophet are then cast alive into a lake of fire, and Satan is locked away for a thousand years in the same bottomless abyss Abaddon had emerged from. Jesus then begins His millennial kingdom and peacefully reigns on Earth for one thousand years with His believers, and with others who survive this terrible time.

(Exodus 33:23, 34:30; Jeremiah 23:6-8, 25:31-33, 31:32-34; Isaiah 9:6-7; 11; 40; Daniel 7:9-14, 12:1-3; Zechariah 14:6-16; Job 19:25-27; Joel 2:1-11; Matthew 7:23, 24:26-39, 25:12,41, 26:64, 27:52-53; Mark 13:25-27; Luke 20:35-36, 24:39; John 6:40, 11:25-26, 20:20-29; Acts 1:9-11, 17:32; 1 Corinthians 15:35-54; 2 Corinthians 5:1-10, 12:2; Colossians 2:18, 3:4; 1 Thessalonians 1:10, 3:13, 4:13-5:3; 2 Thessalonians 2:8; 1 Timothy 4:16, 6:16; 1 John 3:2; Jude 14; Revelation 1:7, 4:1-6, 7:1; 8; 14:15-20, 17:8, 19:10-20:3)

Those who trust in the Lord will be changed
and shall ascend up with wings as eagles.
They will tirelessly run and walk (Isaiah 40:31)

<u>Level Up</u>

Our bodies won't last forever
We are soon to be free
Born on Earth and raised in glory
Pathetic to powerful
Animals to life-giving spirits
Earthly to heavenly
Time to level up

I will raise them at the last day
I am the resurrection, I am the life
Believe in me and live, even after death
Live in me, believe in me and never die
Victory swallows death, where is your sting?

It's by faith, not by sight
To be absent from our body is to be present with our Lord
When He appears like Him we will be, seeing as He is

My Redeemer lives, when on Earth at last He stands
My body decayed yet in my body I see God
I see Him myself and not another, yes, with my own eyes
How my heart yearns

A white cloud, the Son of Man
A golden crown, in His hand a sickle
Says an angel from God's temple
Swing, the time is right, Earth's crop is all ripe
The sickle swings and Earth is harvested
Like a thief in the night
Jesus sends His angels to gather His elect from Earth and the heavens
The Lord descends, a shout, the archangel and trump of God
The dead and living rise, caught up together in the clouds
To meet the Lord in the air
Transformed in a flash at the last trump
When we in immortal bodies be with the Lord forever
Wonderful, Counselor, Mighty God, Everlasting Father
Prince of Peace

"Level Up" is on my *Peace and Power* album,
and as "Superbodies" on my *The Way* album.

The Millennial Reign

As the death of the seed produces a plant, so God will match to each believer's spirit DNA a new spiritual body as different as the sun is from the moon. Our physically weak terrestrial bodies will become powerful new spiritual bodies of majestic splendor, like the resurrected physical body of Jesus, not bound by physics and gravity— When Jesus appears, we will be like Him, because we will see him as He is. We may be able to fly, pass through walls, and teleport.

(Daniel 9:21; Luke 24:31-48; John 12:24; Acts 1:9; 1 Corinthians 15:35-49; I John 3:2)

Believers will have that close personal connection with God and others we so crave at times, but sin prevents us from having. Since we are made in God's image, we will relate to God much more and have a very close personal relationship with Him. We will have already chosen not to rebel against God, but to wholly serve and believe in Him.

The massive earthquake at the end of the Great Tribulation levels Earth's mountains and islands. During Christ's peaceful one-thousand-year millennial reign, the planet may have an incredible interlacing of streams, waterfalls and lakes, with gentle, grass eating lions, leopards, wolves and bears who are all as tame as lambs. Vast giant forests of Jurassic Era redwood and sequoia trees may again populate much of Earth. (Isaiah 11:6-9; Haggai 2:6,21-22; Matthew 26:29; Revelation 16:18-20, 20:4-6)

But, at the end of Christ's one-thousand-year reign, Satan is released and once again turns humans against God and His people. Earth is then destroyed by God in the sixth mass destruction event of fire, and from its ashes God will make a new iEarth 7, beautiful beyond imagination.

God will live with us on a new earth with no sea, and wipe every tear from our eyes and love us. No one will ever die or be depressed. There will be no pain. The way we exist now will be gone forever. God will make everything new for His new earth— new laws of physics, new plants, new trees, new animals, new ways of life, new forms of communication, new ways of transporting ourselves, new ways of feeling, new relationships, new wisdom and understanding of God and our environment, new colors, and delicious new satisfying foods. And God Himself will dwell with us in His fabulous new city.

(Isaiah 11:6-9, 65:18-25; Ezekiel 39; Daniel 2:44-45, 3:25; Amos 9:13-15; Zephaniah 3:8-12; Zechariah 8:4, 14:9-16; Matthew 19:28; John 14:2-4; 2 Peter 3:7-10; Revelation 20)

Father's Brilliant City of Light

An angel took John to an exceedingly high elevation above Earth to view our Father's celestial city descending from the cosmos like a beautiful bride. He heard a loud voice from heaven say, "God's dwelling will now be with humans. He will live with them, and they will be His people. God will be with us and wipe away every tear from our eyes. No longer will there be death or sorrow or pain, because the old way is gone forever." Then God said, "I now make all things new!"

As John watched from an extremely lofty place, an angel measured this radiantly-golden floating city– It is 1,400 by 1,400 miles square. The 1,400-mile *altitude* of her *gates* places her in Medium Earth Orbit (1,150 miles higher than the International Space Station!).

Her vast 200-foot pure jasper wall rests on twelve tremendous foundational jewels. Twelve angels stand by her amazing twelve open gates. Each gate is made of an enormous dazzling pearl that luminously reflects the incredible lights emanating from this fabulous luminescent city— Her mansions and pavement are made of pure, transparent gold!

Since God *is* her temple, His eternal light fantastically radiates within her transparent gold mansions and pavement, and within her red jasper wall and sparkling jeweled foundation, and brilliantly illuminates Earth below! Earth's dynamic green orb mesmerizingly spins beneath the city's transparent foundation, and vast panoramic celestial views of the Universe are visible above through the crystal-clear red jasper walled dome of God's interstellar gateway city. The pure Water of Life River streams from God's throne, and in the open square and on either side of the River is the Tree of Life with its healing leaves, bearing twelve fruits.

Revelation says we will no longer need the sun. God is well aware that our sun will eventually burn out. He knew it long before we figured it out, and He has a plan to address it. God will be our light. God put a little info in there at the end to alleviate the concerns of us future humans who learn about astrophysics and worry about the sun burning out someday. It's all going to be okay. This new heaven, earth and city is God's next step in accomplishing what He wants in His manifold wisdom. Jesus is now preparing this city for us in His Father's house with many mansions. God is love. (Psalms 104:5; Ezekiel 47; Matthew 5:5; John 3:16, 14:2-11; 1 John 4:16-19; Revelation 21-22)

Blessed are the meek, for they shall inherit the earth – Jesus

References

1 Fields, Helen. “Dinosaur Shocker.” *Smithsonian.com*, 1 May 2006, www.smithsonianmag.com/science-nature/dinosaur-shocker-115306469/.

2 Herzing, Denise and Johnson, Christine. *Dolphin Communication and Cognition: Past, Present, and Future* (pp 3-13). The MIT Press. *How Intelligent Are Whales And Dolphins.* us.whales.org/whales-dolphins/how-intelligent-are-whales-and-dolphins/.

3 Stoeckle, Mark Y., and David S. Thaler. “Why Should Mitochondria Define Species?” Human Evolution (Vol. 33 - n. 1-2 (1-30) – 2018). Jones, Andrew. Humans and Animals Are (Mostly) the Same Age? June 8, 2018. evolutionnews.org/2018/06/humans-and-animals-are-mostly-the-same-age/. Sequencing Y Chromosomes Resolves Discrepancy in Time to Common Ancestor of Males V. Females. science.sciencemag.org/content/341/6145/562.

4 www.cslewisinstitute.org/CS_Lewis_and_JRR_Tolkien_page2. *C.S. Lewis and J.R.R. Tolkien – page 2.*

5 Galileo and the Inquisition. galileo.rice.edu/bio/narrative_7.html.

6 Stuart, Rhett. *The Bishop Maker.*

7 *Chris Pratt's 9 Rules Acceptance Speech 2018 MTV Movie & TV Awards.* www.youtube.com/watch?v=EihqXHqxri0.

8 *Chris Pratt thanks Jesus in first appearance since split at Teen Choice Awards.* www.usatoday.com/story/life/entertainthis/2017/08/13/chris-pratt-thanks-jesus-christ-first-appearance-since-split-anna-faris/563664001/.

9 Hawking, Stephen. The Beginning of Time. www.hawking.org.uk/the-beginning-of-time.html.

10 *God, Design, and Fine-Tuning*, Robert Collins citeseerx.ist.psu.edu/viewdoc/download?doi=10.1.1.470.7166&rep=rep1&type=pdf. www.newscientist.com/article/mg20227123-000- gravity-mysteries-why-is-gravity-fine-tuned/

11 Rich Deem. *Quotes from Scientists Regarding Design of the Universe.* godandscience.org/apologetics/quotes.html. Schaefer, Henry. *Scientists and Their Gods.* leaderu.com/offices/schaefer/docs/scientists.html

12 NIH Human Microbiome Project defines normal bacterial makeup of the body. www.nih.gov/news-events/news-releases/nih-human-microbiome-project-defines-normal-bacterial-makeup-body.

13 Key, Alexander. *The Forgotten Door.*

14 Kramarik, Akiane. akiane.com/my-story/. 5/2/2012 Facebook post.

15 *How tech's richest plan to save themselves after the apocalypse* www.theguardian.com/technology/2018/jul/23/tech-industry-wealth-futurism-transhumanism-singularity.

16 bible.org/question/were-three-days-and-three-nights-jesus-was-grave-full-72-hours.

17 Smith, Henry B Jr. Case for the Septuagint’s Chronology in Genesis 5 and 11. biblearchaeology.org/research-articles/the-case-for-the-septuagints-chronology-in-genesis-5-and-11/.

18 Durant, Will & Ariel. *The Lessons of History,* “Morals and History”.

19 A Flat Earth, and other Nonsense. creation.com/refuting-flat-earth.

20 NASA Spacecraft Discovers new Magnetic Process in Turbulent Space. May 9, 2018. www.nasa.gov/feature/goddard/2018/nasa-spacecraft-discovers-new-magnetic-process-in-turbulent-space.

21 Magnetic Reconnection in Earth’s Magnetosphere. sci.esa.int/cluster/51744-magnetic-reconnection-in-earth-s-magnetosphere/.

Websites accessed 11/2019

References

<u>Verse highlights for HEART</u>: HEBREW heart leb (3820) will and intellect: Genesis 6:5, 8:1, 17:17, 42:28, 45:26; Exodus 4:14, 31:6, 35:5-22, 35:26, 36:2, 16:28; Numbers 16:28, 24:13; Deuteronomy 28:65, 29:4; Judges 16:16; 1 Samuel 2:1, 10:9, 27:1; 2 Samuel 6:16, 15:6; 1 Kings 3:12, 4:29, 8:66, 11:3, 18:37, 21:7; 2 Kings 5:26; 1 Chronicles 12:33, 28:9, 29:9; 2 Chronicles 32:26; Psalms 10:17, 12:2, 14:1, 19:14, 33:15, 34:18, 51:10-17, 143:4; Proverbs 1:7, 3:5, 4:23, 18:2, 19:8,21, 20:5, 20:9, 21:4, 22:15,17, 23:19, 28:26, 31:11; Ecclesiastes 1:13-17, 2:15,22, 5:2, 7:3-7, 7:22, 11:10; Song of Songs 8:6; Jeremiah 17:9; Ezekiel 36:26; Daniel 1:8, 10:12; **heart lebab (3824) will, intellect and feelings:** Genesis 20:5, 31:26; Numbers 15:39; Deuteronomy 6:5, 9:4, 11:16, 15:7, 20:3, 28:28,47,67, 29:18; Joshua 2:11, 24:23: 1 Samuel 9:19, 12:24, 13:14, 14:7, 16:7, 17:28, 21:12; 1 Kings 2:4, 8:18,29, 9:4, 10:2, 11:4, 1 Chronicles 29:18; Job 9:4, 34:34; Proverbs 16:23, 25:3; Psalms 4:4, 13:2, 25:17, 28:3, 111:1, 139:23; Isaiah 6:10, 7:2, 10:7, 32:4, 47:8; Zechariah 7:10. **heart mind sekviy (7907)** Job 38:36; **GREEK heart kardia (2588) will, intellect and feelings:** Matthew 5:8, 5:28, 6:21, 11:29, 12:33-34, 13:15, 15:18, 22:37; Mark 2:6, 7:21: Luke 2:19, 6:45, 8:12; 9:47, 16:15, 21:14, 24:25,,32,38; John 14:1, Acts 28:27; Romans 1:24, 5:5, 8:27, 10:9-10, 1 Corinthians 4:5, 14:25, 2 Corinthians 1:22, 2:4, Galatians 4:6, Ephesians 1:18, 3:17, 6:22, Colossians 3:15, Hebrews 3:12, 3:15, 4:12. **sklerokardia (4641) hardness of heart** Matthew 19:8; Mark 10:5, 16:14.

<u>Verse highlights for SPIRIT</u>: HEBREW spirit ruwach (7307): Genesis 1:2, 3:8, 6:17, 8:1, 26:35, 41:8, 45:27; Exodus 6:9, 28:3, 31:3, 35:21,31; Numbers 5:14, 11:17, 11:29, 14:24, 16:22; Joshua 2:11, 3:10, 5:1, 8:3, 9:23; Judges 15:14, 15:19, 1 Samuel 1:15, 16:14, 19:23, 30:12; 2 Samuel 22:11,16; 1 Kings 22:21; 2 Kings 2:9, 2:15; 1 Chronicles 5:26; 2 Chronicles 9:4, 18:20; Ezra 1:5; Nehemiah 9:20; Job 7:11, 10:12, 12:10, 15:13, 17:1, 20:3, 21:4, 32:8,18, 33:4; Psalms 32:2, 33:6, 34:18, 51:10-17, 77:2-6; 78:8, 104:4, 139:7, 142:3; Proverbs 1:23, 11:13, 14:29, 15:4, 15:13, 16:18,19,32, 17:27, 18:14, 29:11; Ecclesiastes 1:14, 4:8, 7:8, 8:8, 11:5; Isaiah 4:4, 11:2, 19:14, 26:9, 29:10, 37:7, 42:1, 48:16, 54:6, 57:15, 61:1,3,10; 66:2, Ezekiel 3:24, 11:19,24, 13:3, 18:31-32, 21:7, 36:26-27, Hosea 5:4; Haggai 1:14; Zechariah 12:1,10, 13:2; Malachi 2:15-16. **GREEK spirit pneuma (4151):** Matthew 1:18, 3:15, 4:1, 5:3, 8:16, 10:1,20, 12:18, 26:41; Mark 1:8,27, 2:8, 5:8; Luke 1:15,47, 2:26,27; John 3:5-7, 14:17-26, 16:13, 19:30; Acts 2:38, 5:3,9,16,32; 6:10, 7:55, 8:39, 9:31, 10:44-45; 11:12, 13:2,52, 15:28, 19:21, 20:28, 21:11; Romans 1:4,9, 2:29, 5:5, 7:6, 8:2-27, 12:11, 14:17, 15:30; 1 Corinthians 2:4,13, 3:16, 6:18-20, 7:34,40, 12:3-7, 16:18; 2 Corinthians 1:22, 2:13, 3:3,6,17, 7:13, 11:14; Galatians 3:2, 4:6, 5:16-25, 6:1,8,18; Ephesians 1:17, 2:2, 2:18,22, 3:16, 4:3,22-23,29,5:18; Ephesians 6:17-18: Philippians 1:27, 3:3; Colossians 1:8, 2:5; 1 Thessalonians 1:5-6, 4:8, 5:19,23; 2 Thessalonians 2:2,8,13, 4:1; 2 Timothy 1:7,14; Titus 3:5; Hebrews 1:14, 4:12, 9:14, 12:9, 23:23; James 2:26, 4:5, 1 Peter 1:12, 3:4,18, 4:6; 1 John 4:13; Jude 19-20.

<u>Verse highlights for SOUL (LIFE)</u>: HEBREW soul naphash (5315): Genesis 1:20-24, 30; 2:7, 9:4, 12:5, 12:13, 14:21, 19:17,20, 27:19, 34:3, 35:18, 42:21; Exodus 15:9; Leviticus 17:14, 18:29; Deuteronomy 11:18, 12:20, 13:4,6, 14:26, 18:6, 23:24, 24:15, 28:65; 1 Samuel 18:1, 23:20; 2 Samuel 17:8; 1 Kings 2:4; 2 Kings 2:2,4,6, 4:30; 1 Chronicles 28:9; Job 10:1, 12:10, 18:4, 21:25, 30:25; Psalms 11:5, 16:10, 29:7, 25:7, 25:13, 27:12, 31:7, 34:2, 35:13, 41:4, 42:1-5, 62:5, 94:19, 103:1, 119:20,25,28,81,167, 139:14, 143:8, 146:1; Proverbs 6:16, 11:17, 11:30, 13:35, 14:10, 16:17, 21:23, 22:23-24, 24:14; Song of Songs 1:7, 3:1-4; Isaiah 10:18, 42:1, 57:16, 61:10; Jeremiah 6:8, 38:16. **GREEK soul psuche (5590):** Matthew 10:28,39, 11:29, 12:18; Mark 8:26, 12:30; Luke 1:46, 12:20,22; John 10:15; Acts 2:41; 1 Corinthians 15:45, Philippians 1:27; Hebrews 4:12, 10:28,30; James 1:21; 1 Peter 1:9, 2:11, 3:20; 2 Peter 2:14, Revelation 6:9.

References

Suggested Reading:
References 1-16, and scripture references given throughout the book are highly recommended.

Books: *Erasing Hell* by Francis Chan & Preston Sprinkle; *Heaven* by Randy Alcorn; *The Case for a Creator / Evidence that Demands a Verdict* by Lee Strobel; *Love Does / Everything Always* by Bob Goff; *Seeking Allah, Finding Jesus* by Nabeel Qureshi; *Mere Christianity* by C.S. Lewis; *The Insanity of God / The Insanity of Obedience* by Nik Ripken; *The Unseen Realm* by Dr. Michael Heiser.

Suggested Study: There are online interlinear Bibles that show the original Greek and Hebrew words with Strongs Concordance numbers. These Strongs numbers can be used for Hebrew (Old Testament) and Greek (New Testament) Englishman Concordance (reverse) word searches to see how they are used in context with other verses throughout the Bible.

About the Author
Rhett Otis and his wife live in Virginia and have four children. He has a Bachelor of Science degree in organizational behavior from California State Polytechnic University, and investigates civil and white-collar fraud. He has believed in Jesus since 1973.

Books
God and Humans
God's Master Plan

Books under pseudonym Rhett Stuart
The Bishop Maker – When the Battle is for the Subconscious
How to recognize and recover from conservative Christian cults
Generation Modified
Young Adult science fiction

Music
Rhett Otis is on most streaming services
Lyrics aspire to scriptural accuracy
Spotify Playlists:
Gospel Indie Favorites
Beach Rock Gospel Indie
Locked In Gospel Indie
Easy Listening Gospel Indie
Country Gospel Indie

Printed in Dunstable, United Kingdom

91230506R00119